Analysis and Visualization of Discrete Data Using Neural Networks

Analysis and Visualization of Discrete Data Using Neural Networks

Koji Koyamada

Kyoto University, Japan

W **World Scientific**

NEW JERSEY · LONDON · SINGAPORE · BEIJING · SHANGHAI · HONG KONG · TAIPEI · CHENNAI · TOKYO

Published by

World Scientific Publishing Co. Pte. Ltd.

5 Toh Tuck Link, Singapore 596224

USA office: 27 Warren Street, Suite 401-402, Hackensack, NJ 07601

UK office: 57 Shelton Street, Covent Garden, London WC2H 9HE

Library of Congress Cataloging-in-Publication Data

Names: Koyamada, Kōji, author.

Title: Analysis and visualization of discrete data using neural networks / Koji Koyamada.

Description: New Jersey : World Scientific, [2024] | Includes bibliographical references.

Identifiers: LCCN 2023038211 | ISBN 9789811283611 (hardcover) |
    ISBN 9789811283628 (ebook for institutions) | ISBN 9789811283635 (ebook for individuals)

Subjects: LCSH: Data mining. | Information visualization--Data processing. |
    Neural networks (Computer science)

Classification: LCC QA76.9.D343 K685 2024 | DDC 006.3/12--dc23/eng/20231222

LC record available at https://lccn.loc.gov/2023038211

British Library Cataloguing-in-Publication Data

A catalogue record for this book is available from the British Library.

For any available supplementary material, please visit
https://www.worldscientific.com/worldscibooks/10.1142/13603#t=suppl

Desk Editor: Quek ZhiQin, Vanessa

Typeset by Stallion Press
Email: enquiries@stallionpress.com

Contents

Chapter 1

Introduction

In this chapter, you will learn practical methods for analyzing physical data spatiotemporally defined by discrete points. Physical data refer to data used to describe physical or other phenomena. The physical phenomena refer to various phenomena that exist in nature. They include, for example, weather, temperature, earthquakes, and floods. The physical data are often defined as spatiotemporal discrete data. The spatiotemporal discrete data refer to data that are not spatiotemporally continuous. When we handle spatiotemporal discrete data, we usually divide the spatiotemporal axis into discrete segments so that we can handle data in each segment.

Spatiotemporal discrete data can be used in a variety of fields. For example, in the social and natural sciences, spatiotemporal discrete data are sometimes used to analyze various phenomena and their continuous changes. In biology and geology, spatiotemporal discrete data are sometimes used to analyze phenomena that occur in time.

Physical data can be measured in a variety of ways as follows:

- *Using measuring instruments*: Physical data can be measured directly using measuring instruments. For example, a thermometer or a hygrometer can be used to measure temperature and a seismometer can be used to measure the vibrations of an earthquake.
- *Observation*: Physical data can be obtained by observing phenomena. For example, a weather forecast site or app can be used to observe the weather. To acquire weather data, you can observe temperature, humidity, wind speed, etc., at meteorological observatories.
- *Using simulations*: Physical data can be obtained by running simulations on a computer. For example, you can understand fluid and heat conduction phenomena by simulating fluid behavior and heat conduction phenomena.

Today, discovering a spatiotemporal model that explains physical data defined by spatiotemporal discrete points (hereafter simply referred to as "physical data") is an interesting topic for researchers and practitioners. A key topic in the discovery of a spatiotemporal model is the discrete-to-continuous transformation to effectively evaluate partial differential terms. More specifically, it is about finding a partial differentiable function that adequately approximates physical data defined discretely in the time space. A partially differentiable function is characterized by a partial derivative at a given point. Partial differentiation refers to calculating a derivative with respect to a particular variable in a multivariable function at that point. The partial differentiable function, which has a partial derivative at a given point, is used to analyze a multivariable function. Successfully finding such partial differentiable function enables effective visualization and also enables a partial differential equation (PDE) that explains a phenomenon to be derived using the results of analysis.

This document uses Excel to understand the basics of how to analyze physical data. Excel is widely used as a piece of spreadsheet software to organize data is suited for data analysis. Using Excel has the following benefits:

- *Easy to use*: Excel is commonly pre-installed on most computers and does not require additional installation. Moreover, Excel can be used without the need for specialized skills.
- *Has a variety of functions*: Excel has a variety of functions that allow you to aggregate data efficiently.
- *Can create graphs or charts*: Excel can be used to visually display data in the form of a graph or chart, which facilitates understanding of data.
- *Data can be easily imported*: Excel allows you to import data from, for example, a CSV or text file.
- *Can work jointly*: Excel allows multiple people to work jointly, making it useful for data analysis projects.

Excel, which provides an array of benefits, is also widely used in data analysis. Next, the basics of Excel will be explained.

1.1. Basic operations of Excel

Excel is a piece of spreadsheet software developed by Microsoft. The spreadsheet software is a tool that allows you to enter and edit data in a tabular format for aggregation, analysis, and other purposes. Excel is widely used by individuals and in businesses, and can handle large amounts of data, making it suitable for aggregation and analysis. Excel can be used to not only create a table but to also create a graph or chart. Additionally, Excel has a variety of functions to organize data, allowing you to analyze data and calculate formulas. Excel is commonly pre-installed on Windows and Mac OS and sold as part of Microsoft Office. It is also available as a cloud service, so you can use it from any Internet-connected environment. The data to be analyzed are often provided in Excel format. There are quite a few tools capable of reading and analyzing Excel data, but Excel itself also contains functions to perform analyses. Excel provides tools for determining correlation between data, testing for differences in the mean and variance of data, and regression analysis. It also serves as an optimization tool. These tools can be used to implement deep learning. Excel can be regarded as a tool for preliminary analysis prior to utilizing dedicated analysis tools, and it is advisable to become familiar with its functionalities. This section describes the basic terminology in Excel and explains how to perform what-if analysis using sample data.

1.1.1. *Table components*

The main components of an Excel table are as follows (Figure 1.1):

- *Worksheet*: One of the basic building blocks of an Excel document. One document has multiple worksheets. One worksheet is usually contained in an Excel file, but multiple worksheets may be included.
- *Cell*: A basic element for entering data in Excel at which a row and a column intersect. You can enter text, numbers, and other values in a cell.
- *Row*: A basic element that divides an Excel worksheet horizontally. A row usually contains a single cell. In Excel, you can refer to a row by specifying a row number.
- *Column*: A basic element that divides an Excel worksheet vertically. A column usually contains a single cell. In Excel, you can refer to a column by specifying the column number.
- *Row number*: A number that identifies a row in Excel and appears on the left. Row numbers typically start from "1".
- *Column number*: A number that identifies a column in Excel and appears at the top. Column numbers usually start from "A".

Each component of a table in Excel plays an important role in creating and manipulating a table. When creating a table or entering data, it is

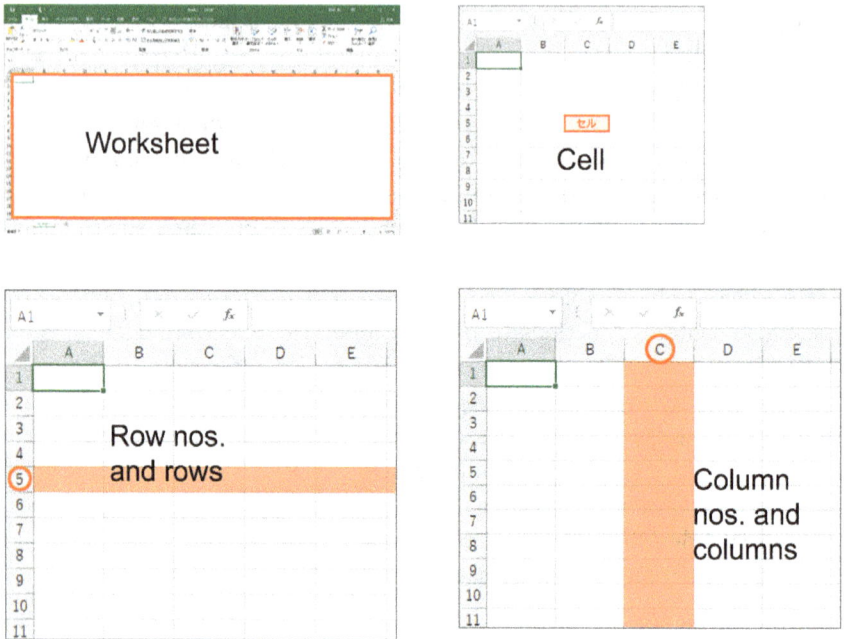

Figure 1.1. Components of a table in Excel.

important to understand these components and manipulate them appropriately.

1.1.2. *Name box and formula bar*

The name box and formula bar (Figure 1.2) in Excel have the following roles:

- Name box: In Excel, you can name a specified cell or range. This is called naming. Naming allows you to specify a cell or a range of cells. A named cell or range is called a "name" in Excel. You can manage your data more efficiently by specifying a named cell or range. Excel provides a function called "name box" that allows you to specify a named cell or range by entering its name.
- Formula bar: In Excel, you can calculate data by entering a formula. You can enter a formula in a place called "formula bar". The formula bar is located in the Excel status bar and used to enter a formula into a cell. Entering a formula allows you to calculate data efficiently.

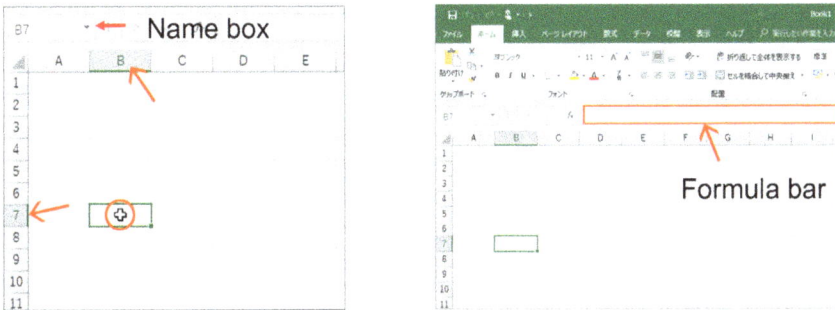

Figure 1.2. Name box and formula bar.

Figure 1.3. Ribbon.

1.1.3. *Ribbon*

The Excel ribbon refers to the tabbed menus that appear in the main Excel window. The ribbon has various functions and commands grouped together into a tab according to the category. There are various tabs, such as Home and Data. Users can use commands in the ribbon to manipulate Excel to create or edit a table.

The tabs and the area containing the buttons which can be switched by clicking a tab are collectively called the "ribbon" (Figure 1.3). You can switch between buttons by clicking the tab placed above them.

1.1.4. *File tab and Backstage view*

The Excel file tab is a special tab that appears in the upper-left corner of the main Excel window. Clicking on the File tab will bring up the Backstage view (Figure 1.4).

The Backstage view is a space for managing Excel files and data. The Backstage view contains templates for creating a new workbook, as well as

File tab

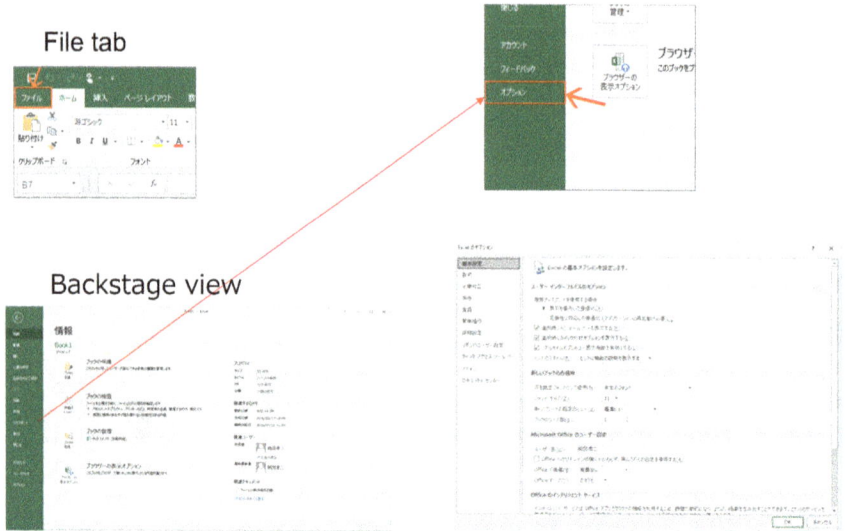

Backstage view

Figure 1.4. File tab and Backstage view.

commands for opening, saving, printing, and sharing existing workbooks, and for setting Excel options. If you want an Excel data analysis tool or solver to appear as a menu on the ribbon, you must set a corresponding option here. In the Backstage view, you can also check your Excel account or privacy settings.

Clicking on the File tab will bring up the Backstage view. Clicking Options will display a dialog box.

1.1.5. *Autofill*

The Autofill function in Excel helps you enter data efficiently. It automatically suggests data to be entered next based on the data that have already been entered (Figure 1.5).

For example, suppose that columns contain month and day data. The data "January", "February", and "March" are entered in the "Month" column, and the data "1st", "2nd", and "3rd" are entered in the "Day" column. If you enter "4" in the April cell of the "Month" column, "4" will also be automatically entered in the 4th cell of the "Day" column.

The Autofill function allows you to enter data more efficiently, so you can perform Excel work more efficiently. The Autofill function also reduces input errors, thereby increasing data reliability.

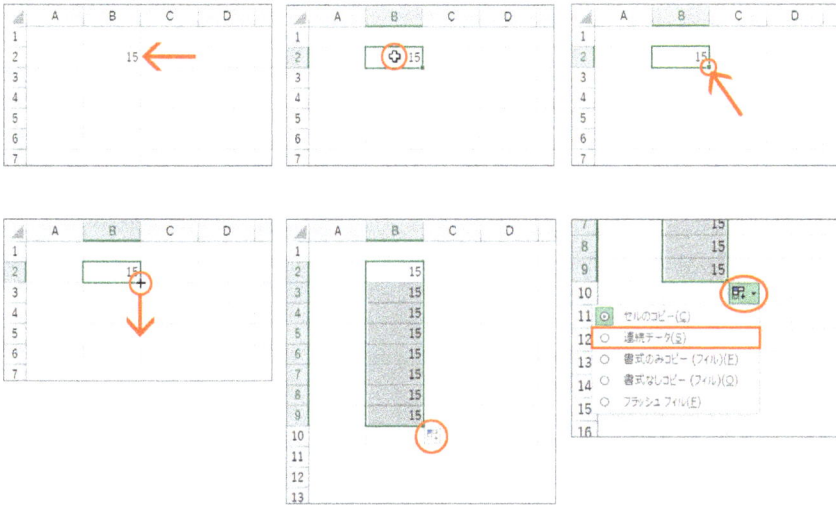

Figure 1.5. Autofill.

You can easily enter continuous data such as "January, February, March..." by dragging them:

1. Select a cell filled with a number.
2. Move the mouse pointer to the small square at the bottom right of the thick frame.
3. When the shape of the mouse pointer changes into a cross, drag it.
4. Click on "Autofill Options" that has appeared in the lower right area.

1.1.6. *Relative reference*

In Excel, the relative reference function allows the reference destination in a referenced cell to be automatically changed when the referenced cell is moved (Figure 1.6).

For example, for a cell containing the formula "= A1", if you were to move this cell to the right by one cell, the formula will be automatically rewritten to "= B1". The function capable of changing the reference destination in a cell automatically when the cell is moved is known as "relative reference".

The "relative reference" function, which automatically changes the reference destination in a cell when the cell is moved, is useful for writing formulas. When data that use "relative reference" are duplicated, a reference

1. Goal   2. Select D2.   3. Drag downwards.

4. Check that Autofill   5. Select D3.   6. Checking cell
works properly.   contents

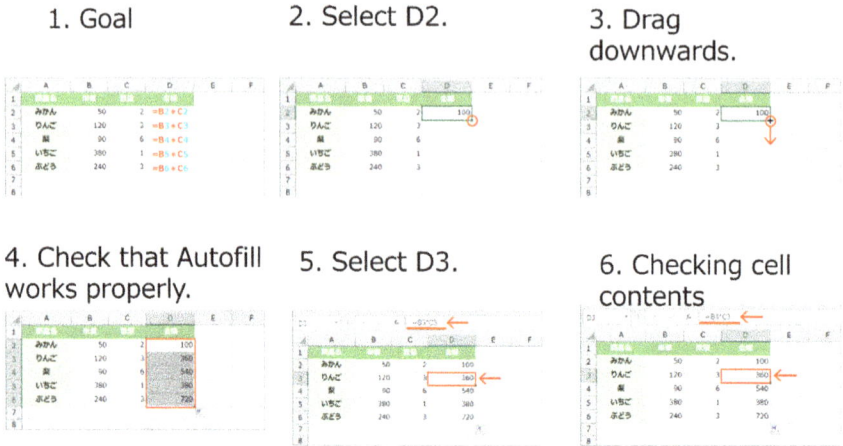

Figure 1.6. Relative reference.

destination in the duplicated cell is also changed, allowing you to manage data efficiently.

Autofill is used to copy a formula.

1. If you select the D2 address that contains a formula, a small square will appear at the bottom right of that cell.
2. Drag it downward to automatically fill other cells.
3. Select the B2 address. The contents of that cell will be displayed in the formula bar.

1.1.7. *Absolute reference*

In Excel, when a cell containing "absolute reference" is referenced, the reference destination in the cell is fixed even after the cell moves.

For instance, consider a cell that contains the formula = "$A$1". If you shift this cell to the right by one cell, the reference destination in the moved cell will remain unchanged as "=$A$1". The function capable of fixing a reference destination in a cell even after the cell is moved is known as "absolute reference".

The absolute reference function, which prevents the reference destination in a cell from being changed after the cell has moved, is useful for writing formulas. When data that use "absolute reference" are duplicated, the reference destination in the duplicated cell is not changed, allowing you to manage data efficiently (Figure 1.7).

1. Check cell contents.

2. Choose B4.

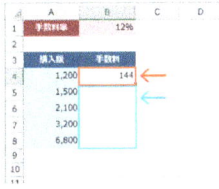

3. Drag downwards.

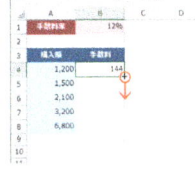

4. Autofill fails.

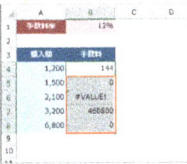

5. Click B4 to select B1

6. Press "F4."

B1 → $B$1 → B$1 → $B1

Figure 1.7. Absolute reference.

It is recommended to use the absolute reference function in order to retain the cell address in a formula when the formula is copied.

1. Select the B4 address. This will cause the cell to be surrounded by a thick frame with a small square appearing at the bottom right.
2. Dragging the small square downward will not trigger the autofill to work.
3. Instead, click to select B1, which is the "commission rate", and press F4.
4. Make sure the column and row numbers are prefixed with "$" to indicate they are absolute references. There are three prefixing patterns: absolute reference for both column and row, for row only, and for column only. This pattern can be changed by pressing the F4 button.

To learn how to use relative and absolute references, display the calculation results of numbers in column A × numbers in row 1.

1. Enter 1 to 9 downward, starting from row 2 of column A.
2. Enter 1 to 9 rightward, starting from row 1 of column B.
3. Calculate products for all 9 × 9 cells using relative and absolute references appropriately (Figure 1.8). More specifically, enter a formula "= $A2*B$1" in row 2 of column B. Copy this cell up to row 10, and then copy all the rows 2 to 10 of column B together up to 10 columns.

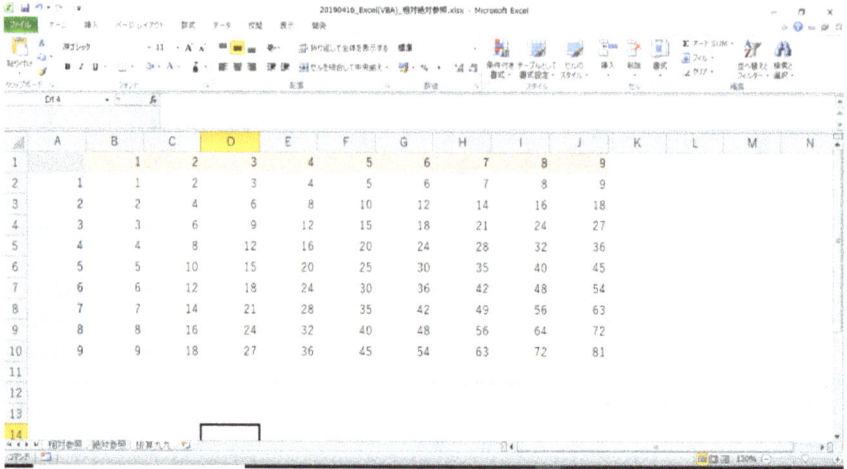

Figure 1.8. Creation of a 9 × 9 multiplication table.

1.1.8. *Introduction of visualization with Excel*

Excel is widely used as a piece of spreadsheet software but it is also suitable for visualizing data.

There are several ways to visualize data in Excel:

1. Visualization with charts: Excel provides a variety of chart types to help you visualize your data with bar, line, pie charts, etc.
2. Visualization using a pivot table: You can visualize data by using an Excel pivot table to aggregate data and convert them to a chart.
3. Visualization using a spreadsheet: You can use an Excel spreadsheet to organize data and visualize them using color coding or formulas.
4. Visualization using macro: Excel has a macro language called VBA. You can use it to create a custom chart or automated visualization tool.

Before visualizing data in Excel, you may have to organize and process them. Excel also has a visualization add-in that allows you to do more detailed visualizations.

Displaying a chart using Excel:

1. Creating a chart: Excel has a variety of chart types that allow you to enter data and create a chart.

   - Select data, select Chart from the Insert tab, and select a chart type.
   - After a chart is created, the data appear in the chart area.

2. Editing a chart: You can edit a chart you have created.

   - Select a chart and select Chart Options from the Design tab.
   - You can change detailed chart settings by using tabs, such as Axis, Data Series, and Layout.

3. Chart design: You can change the appearance of a chart.

   - You can change the appearance of a chart by selecting a preferred chart and choosing Shape, Frame, etc., from the Format tab.

A concrete example is used to explain this. The following is a two-dimensional function: $f(x, y) = \sin(x) \cdot \cos(2y)$. Calculate $f(x, y)$ using a variable value (y) in column A and variable value (x) in row 1 (Figure 1.9).

- Autofill is used to fill variable values downward from the cell of row 2 of column A.
- Autofill is used to fill variable values rightward from the cell of row 1 of column B.

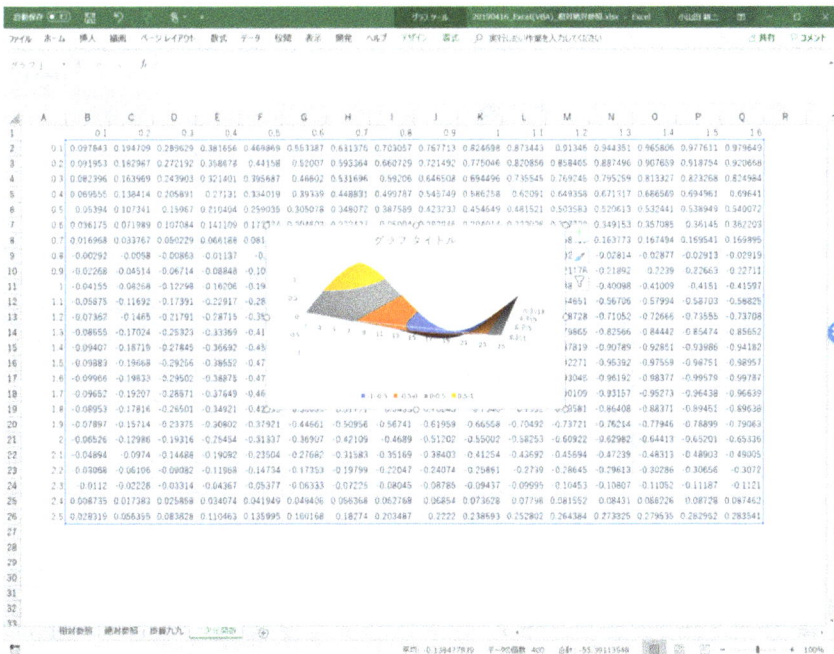

Figure 1.9. Presentation of two-dimensional function as a contour chart.

- Calculate function values using relative and absolute references as appropriate.
- Set the chart type to "Contour" and display a chart.

1.1.9. *Introduction of PDE derivation with Excel*

This section introduces how to use data analysis (regression analysis) in Excel to derive a PDE from the given data; detailed operations will be described in subsequent sections. Regression analysis is a method of analyzing the degree of impact (weight) of two or more explanatory variables on an objective variable. As a concrete example, consider a model that describes the annual sales of a restaurant. Each restaurant corresponds to a row in Excel. In each row, enter annual sales (unit: 10,000 yen), the number of seats (number), walking time from the nearest station (minutes), and presence/absence of breakfast $(0, 1)$. The regression analysis tool allows you to easily create a formula that explains annual sales by adding a weight to each of the following explanatory variables, and an intercept: Number of seats (number), walking time from the nearest station (minutes), and presence/absence of breakfast $(0, 1)$ (Figure 1.10).

This is an introductory exercise for general regression analysis techniques, but it provides a tip for developing a data-driven method for deriving a PDE (Figure 1.11). For example, enter the partial derivatives,

$$u_t, u_x, u_y u_{xx}, u_{yy}$$

Store	AS: Annual sales (10,000 yen)	NS: Number of seats	WT: Walking time from station (minutes)	B: Presence or absence of breakfast (Yes = 1, No = 0)	NP: Number of store limited products (pieces)
A	7800	54	6	0	2
B	8718	75	3	1	4
C	9401	80	1.5	1	5
D	8596	85	4	1	1
E	7235	40	7	0	0
F	9396	76	1.5	1	6
G	7749	45	9	0	2
H	9288	70	2	1	6
I	7581	50	7	0	1
J	8434	78	8	1	4

AS=25*NS-100*WT+350*B+7400

These coefficients are estimated by using the regression analysis

Figure 1.10. Explanation of sales using regression analysis.

座標	u_t	u_x	u_y	u_{xx}	u_{yy}
A点	0.666317411	0.577286767	0.693289572	0.000540143	0.859483388
B点	0.551280683	0.618412384	0.535038373	0.689154679	0.226394308
C点	0.739602382	0.678568024	0.360924624	0.63289503	0.832931637
D点	0.7198427	0.674589904	0.677126491	0.740916367	0.817325912
E点	0.451644194	0.450134344	0.171568854	0.875120374	0.391941881
F点	0.319076964	0.730460862	0.502043833	0.418780668	0.087613776
G点	0.05224234	0.078341697	0.581156924	0.081713198	0.056670549
H点	0.757619307	0.749410916	0.652684555	0.631038237	0.956615967
I点	0.902915453	0.269017146	0.57524135	0.040082549	0.818018695
J点	0.801579605	0.526665113	0.643173508	0.684133799	0.660314115

$$u_t = 0.026u_x + 0.075u_y + 0.116u_{xx} + 0.629u_{yy}$$

These values are estimated using multiple regression analysis.

Figure 1.11. Derivation of a PDE using regression analysis.

for the physical data (u) at each of the multiple coordinate points in each Excel row. The regression analysis tool allows you to analyze the degree of impact of each partial derivative on u [1]. As a result, a PDE

$$u_t = a_0 u_x + a_1 u_y + a_2 u_{xx} + a_3 u_{yy}$$

is derived that describes the physical data.

As a concrete example, consider an equation of motion (Figure 1.12). An equation of motion is an ordinary differential equation representing the laws of motion of an object in physics. A schematic diagram is shown here where the mass of an object is m kg, a spring constant is k, and the proportional coefficient of air resistance is c.

This ordinary differential equation is solved to calculate the position coordinate of the object while changing the time data. This equation becomes a second-order ordinary differential equation at the object's position coordinate x, which can be solved by giving an initial condition. Enter the coordinate data (x) and the differential value $x_t, x_{tt}, u_y u_{xx}, u_{yy}$ at each of the multiple times in each Excel row. x_{tt}, x_t, x regression analysis tools allow you to analyze the degree of impact of position and speed on x_{tt}. As a result, a PDE

$$x_{tt} = a_0 x_t + a_1 x + a_2 \tag{1.1}$$

is derived that describes the physical data.

The advection equation is a PDE that describes the phenomena in which physical quantities, such as a substance and momentum, are carried along

$$ma = -cv - kx$$

$$x_{tt} + \frac{c}{m}x_t + \frac{k}{m}x = 0$$

$$c = 4m,\ k = 5m$$

$$x_{tt} + 4x_t + 5x = 0$$

$$k \qquad x(t)$$

$$x = \exp(-2t)\cos(t)$$

$$\left(\begin{array}{l} x_{tt} = Ax_t + Bx + C \\ A = -4.006603223 \\ B = -5.0502897 \\ C = 1.11022E-16 \end{array} \right)$$

Time t	Position x	Velocity (first derivative)	Acceleration (second derivative)
0.1	0.814641		
0.2	0.656958	−1.45170364	
0.3	0.5243	−1.215494456	2.222138219
0.4	0.413859	−1.007275996	1.945645476
0.5	0.322845	−0.826365361	1.68045945
0.6	0.248586	−0.671184106	1.43373552
0.7	0.188608	−0.539618257	1.209512414
0.8	0.140663	−0.429281623	1.009573958
0.9	0.102751	−0.337703465	0.834119271
1	0.073122	−0.262457769	0.682277033
1.1	0.05026	−0.201248059	0.552494152
1.2	0.032872	−0.151958939	0.442824263
1.3	0.019868	−0.112683206	0.351137261
1.4	0.010336	−0.081731486	0.275267294
1.5	0.003522	−0.057629747	0.213113432
1.6	−0.00119	−0.0391088	0.162704485
1.7	−0.0043	−0.02508885	0.122237123
1.8	−0.00621	−0.014661375	0.090094547
1.9	−0.00723	−0.007069941	0.054274673
2	−0.00762	−0.003806441	−0.003721022

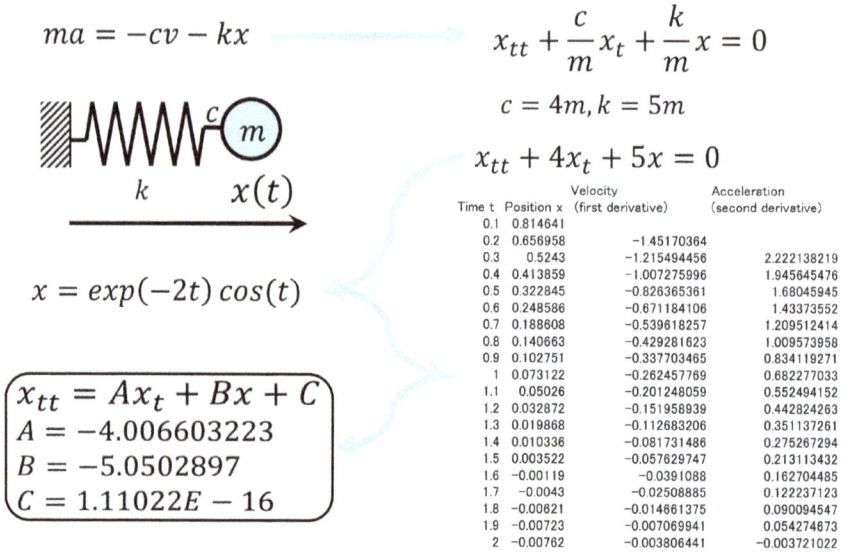

Figure 1.12. Derivation of equation of motion.

with a flow. As another example, consider a case (translational equation) where a substance is transported in a flow at a constant velocity a_0 and the concentration of the substance is $C : u(x, t)$. Figure 1.13 shows a schematic diagram for a one-dimensional equation.

The advection equation given by

$$u_t + a_0 u_x = 0 \tag{1.2}$$

has an exact solution in which the geometry given by initial conditions moves along the x axis at a velocity of a_0.

The concentration u can be calculated by setting the time t and the position x and using this exact solution. Furthermore, this exact solution itself can be differentiated to calculate time derivative of uu_t or space derivation of uu_x. Regression analysis is then used to obtain a regression equation that describes time derivative as spatial derivative. This regression equation is a derived PDE (Figure 1.13).

As described earlier, if the physical data are obtained and a partial derivative can be obtained for each piece of data, a PDE can be derived in a data-driven manner using an Excel analysis tool. The above example deals with a PDE consisting of linear terms where a constant coefficient

$u(x,0) = 1 - x^2$

$$u_t + a_0 u_x = 0$$

$$\frac{dx}{dt} = a_0 > 0$$

$$\zeta = x - a_0 t$$

$$u(x,t) = \begin{cases} 1 - \zeta^2 & |\zeta| < 1 \\ 0 & |\zeta| \geq 1 \end{cases}$$

$$u_t = A u_x + B$$

$$B = 0$$

$$A = -1$$

u	x	t	ux	ut
0	-2.4	0	0	0
0	-2.2	0	0	0
0	-2	0	0	0
0	-1.8	0	0	0
0	-1.6	0	0	0
0	-1.4	0	0	0
0	-1.2	0	0	0
0	-1	0	2	-2
0.36	-0.8	0	1.6	-1.6
0.64	-0.6	0	1.2	-1.2
0.84	-0.4	0	0.8	-0.8
0.96	-0.2	0	0.4	-0.4
1	0	0	0	0
0.96	0.2	0	-0.4	0.4
0.84	0.4	0	-0.8	0.8
0.64	0.6	0	-1.2	1.2
0.36	0.8	0	-1.6	1.6
0	1	0	-2	2
0	1.2	0	0	0
0	1.4	0	0	0
0	1.6	0	0	0
0	1.8	0	0	0
0	2	0	0	0
0	2.2	0	0	0
0	2.4	0	0	0
0	2.6	0	0	0

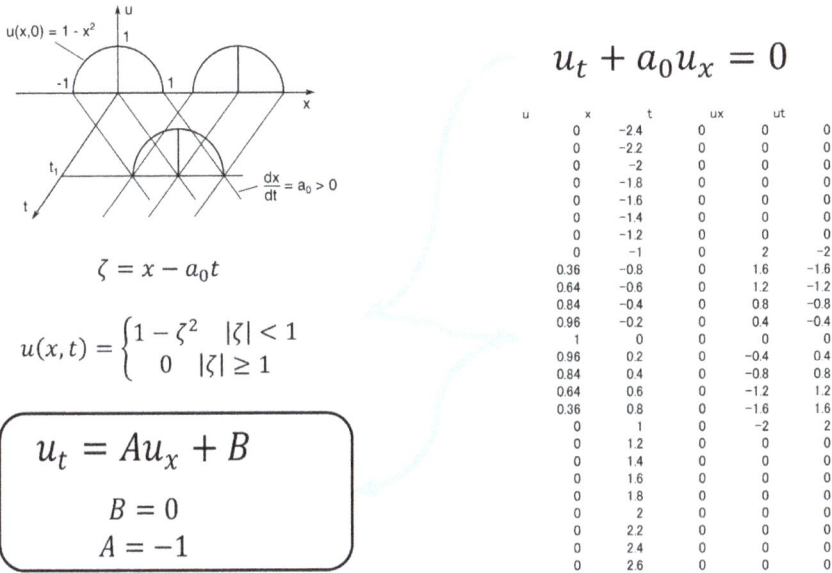

Figure 1.13. Derivation of advection equation.

is assumed for partial differential terms. Even for nonlinear terms where a coefficient is a function of physical data, they can be reduced to a regression analysis problem by integrating the nonlinear parts with partial differential terms.

For the partial differential terms, if an exact solution is given, an exact differential value can be obtained. If not, the calculation of a differential value becomes an issue. One possible strategy is to obtain and differentiate an approximate function that distributes physical data continuously in some manner. One such approximate function that attracts attention is an approximate function based on neural network (NN). This will be discussed in Chapter 2.

1.2. Basic operations of Google Colab (Colab)

Approximately 1000 rows of data can be processed with Excel; however, in order to understand data analysis techniques handling a large amount of data, Google Colab is used. Google Colab (Colab) is Google's cloud environment for data analysis [2]. Colab allows you to start analyzing data immediately on your browser without requiring hardware or software. Data analysis in Colab has the following advantages:

- Eliminates the need to set up hardware or software: Colab allows you to start analyzing right away and proceed with analysis without having to worry about preparation on a PC or server or installation of necessary tools.
- Available in a browser so you can access it from any device: Colab runs in a browser so you can access it from any device, including your PC, smartphone, and tablet.
- Works with a large amount of data: Colab, which works in Google's cloud infrastructure, can handle large amounts of data.
- Keeps a history of your work as you can proceed with work in a notebook format: Colab allows you to proceed with data analysis in a notebook format, so you can keep a history of your work. This enhances repeatability of your work and facilitates collaboration.

Google Colaboratory (Colab for short) is a free cloud service provided by Google, where you can use the Jupyter notebook and Python to analyze data and build a model for machine learning. The following are some basic Colab operations:

1. Creating a notebook.
2. Opening a notebook.
3. Saving a notebook.
4. Sharing a notebook.
5. Connecting a notebook.
6. Running a notebook.
7. Downloading a notebook.

A Colab notebook has a "cell" in which to write text and code. A notebook is composed of multiple cells that can be edited directly in a web browser. The Colab notebook has two types of cell:

- Code cell: A cell for writing Python code.
- Text cell: A cell for describing text that can use Markdown notation to decorate your text (Figure 1.14).

To create a new cell, click "+ Code" or "+ Text" in the menu bar. Alternatively, you can also copy and paste existing cells. To run each cell, click to select it and then press "Shift + Enter" or click the "▶" Run button in the cell.

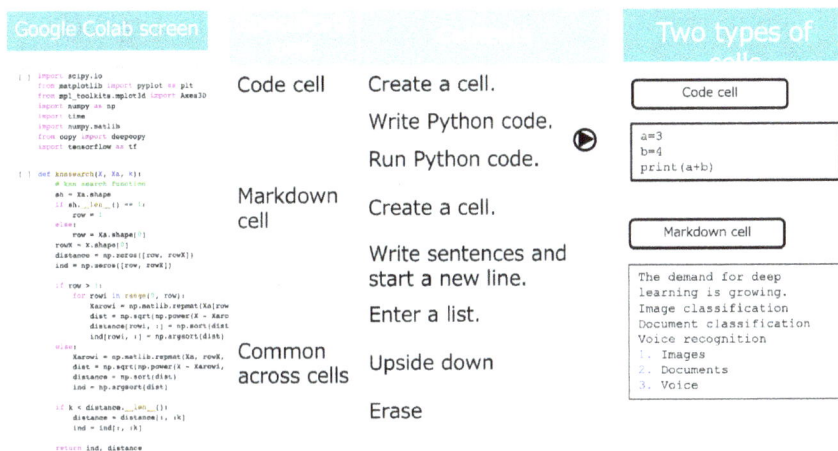

Figure 1.14. Google Colab.

1.2.1. *Code cell*

You can enter and execute a Python code in a code cell.

- To create a code cell, click "+Cell" at the top of a Colab page.
- To enter and run code, select a cell and click the Run button.

Colab has Python libraries pre-installed, allowing you to program with popular libraries like TensorFlow [3], Keras [4], and Numpy [5]. Colab can also perform high-speed computation using GPU and can perform deep learning tasks such as machine learning and image recognition. Colab also has other functions, such as importing external data, saving them in Google Drive, and publishing them to GitHub.

1.2.2. *Text cell*

The text cell allows you to write program descriptions, formulas, links, images, and more.

- To create a text cell, click "+ Cell" at the top of a Colab page and select "Text".
- To enter text, select a cell and enter a sentence in it.

You can use the Markdown notation to format headings, lists, and more.

Colab text cells are very useful for writing descriptions and documents. In particular, they can be used to describe programs and interpret results.

The Colab text cells also allow you to better describe the output of your program.

1.2.3. *Introduction to visualization with Colab*

There are many ways for visualization with Colab.

1. Visualization with Matplotlib [6]: Matplotlib, pre-installed in Colab, is the most commonly used Python library for data visualization. Matplotlib allows you to draw a variety of charts, such as line, scatter, and bar charts.
2. Visualization with Seaborn [7]: Seaborn is a data visualization library based on Matplotlib, allowing you to draw a beautiful chart with ease.
3. Visualization with Plotly [8]: Plotly is a Python library that can be used to create an interactive chart. Plotly allows you to draw a beautiful and interactive chart.
4. Map visualization with Folium [9]: Folium is a library for map visualization in Python that allows you to display markers, polygons, etc., on a map.

For example, you can write the following code to draw a line chart with Matplotlib:

```
< Start >
import matplotlib.pyplot as plt
# X-axis value
x = [1, 2, 3, 4, 5]
# Y-axis value
y = [10, 20, 30, 40, 50]
# Draw a line chart
plt.plot(x, y)
# Chart title
plt.title("Line Plot")
# X-axis label
plt.xlabel("X-axis")
# Y-axis label
plt.ylabel("Y-axis")
# Display chart
plt.show()
< End >
```

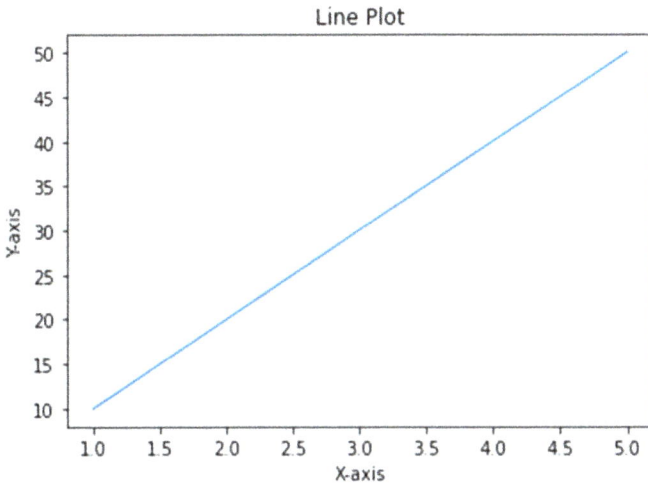

Figure 1.15. Example of drawing a line chart using Matplotlib.

The code above stores the X axis and Y axis values in x and y, respectively, and draws a line chart using the **plt.plot** () function. The **plt.title** (), **plt.xlabel** (), and **plt.yllabel** () functions are used to set the title of the chart, the X-axis label, and the Y-axis label, respectively. Finally, the **plt.show** () function is used to display the chart. As described above, Matplotlib allows you to draw a line chart easily (Figure 1.15).

1.2.4. *Introduction of deep learning with Colab*

To conduct deep learning on Colab, you can perform the following steps:

1. Create a Colab notebook.
2. Install the libraries required for your notebook, such as TensorFlow and Keras.
3. Upload the data required for learning to Colab.
4. Build an NN used for learning.
5. Train the NN with learning data.
6. Make predictions using the trained model.

Colab has pre-installed libraries required for deep learning, such as Ten-sorFlow and Keras, so deep learning can commence immediately just after importing the libraries. Colab also offers GPUs for free so that high-speed training using large amounts of data can be conducted. It is worth noting that Colab is a cloud service provided by Google; therefore, it must be utilized with connection to the Internet, and data storage is the user's responsibility.

As a sample, we will introduce a TensorFlow program that trains the NN to recognize handwritten digits using the Modified National Institute of Standards and Technology (MNIST) handwritten digit dataset. MNIST is a modified version of the original dataset provided by the National Institute of Standards and Technology. The MNIST handwritten digits dataset is an image dataset of handwritten digits widely used in the field of machine learning and deep learning. This dataset contains handwritten digits from 0 to 9, each being a grayscale image of 28×28 pixels. The dataset contains 60,000 images for training and 10,000 for testing. MNIST is widely used as a benchmark for developing and evaluating machine learning algorithms and models.

```
< Start >
# Importing Tensorflow and MNIST
import tensorflow as tf
from tensorflow import keras
# Reading MNIST data
(x_train, y_train), (x_test, y_test) = keras.datasets.mnist.load_data()
# Building a model
model = keras.Sequential([
    keras.layers.Flatten(input_shape=(28, 28)),
    keras.layers.Dense(128, activation='relu'),
    keras.layers.Dense(10, activation='softmax')
])
# Compiling the model
model.compile(optimizer='adam',
      loss='sparse_categorical_crossentropy',
      metrics=['accuracy'])
# Training
model.fit(x_train, y_train, epochs=5)
# Evaluation
test_loss, test_acc = model.evaluate(x_test, y_test, verbose=2)
print('\nTest accuracy:', test_acc)
< End >
```

The above code uses Tensorflow's keras APIs to read MNIST data and build, train, and evaluate an NN.

MNIST data reading: **keras.datasets.mnist.load_data()** is a Keras function that downloads and reads the MNIST dataset. MNIST is an image dataset of handwritten digits, consisting of 60,000 training images and 10,000 test images.

(**x_train, y_train), (x_test, y_test**) are the types of variables to receive four NumPy arrays returned by a function. **x_train** and **y_train** are NumPy arrays representing training images and their corresponding labels (numbers). Similarly, **x_test** and **y_test** represent test images and their corresponding labels.

More specifically, **x_train** is a NumPy array containing 60,000 elements, each representing a grayscale image of 28×28 pixels and being an integer value between 0 and 255. **y_train** is a NumPy array containing 60,000 elements, each being an integer between 0 and 9 and representing a number written in the corresponding image. Therefore, **x_test** and **y_test** are similar, in that each NumPy array contains 10,000 pieces of test data.

These arrays are used to define and train a model in Keras.

In model building, Keras is used to define an NN model and each instruction in the above coding has the following meaning:

- **keras.Sequential ()**: A class for creating a Sequential model. This is a simple model that can be stacked linearly.
- **keras.layers.Flatten(input_shape=(28, 28))**: Defines a flatten layer that converts a two-dimensional array of 28×28 pixels into a one-dimensional flat array as an input layer. This means you can arrange the images in a row and treat them as input.
- **keras.layers.Dense(128, activation='relu')**: Defines dense layers, each having 128 neurons. These are fully bonded layers in which all units in the previous layer are connected to all units in the next layer. The ReLU activation function is applied to these layers to introduce nonlinearity, allowing the model to learn more complex functions.
- **keras.layers.Dense(10, activation='softmax')**: Defines a dense layer having 10 neurons. This is a final output layer which outputs a probability distribution for 10 classes. These probabilities are calculated by the softmax function.

Stacking these layers on the Sequential model defines the NN from the input layer to the hidden layer to the output layer. The model receives a handwritten digit image of 28×28 pixels as input and is trained to classify it into one of 10 classes.

In model compiling, Keras is used to compile the NN model and each instruction in the above coding has the following meaning:

- **optimizer='adam'**: Specifies that the Adam optimizer is used as an optimization algorithm to update model parameters. Adam is one type of optimization algorithm based on the gradient descent method and is characterized by fast speed and fast convergence.
- **loss='sparse_categorical_crossentropy'**: Sets a loss function of the model. In this case, a sparse version of the cross-entropy loss is used for classification tasks. This version is used when the correct answer labels are each indicated as a scalar value and encoded as a class index.
- **metrics=['accuracy']**: Sets metrics to evaluate the model. In this case, this specifies that the correct answer rate is used to evaluate the model's performance. The correct answer rate represents the percentage of samples for which prediction matches the correct answer.

These settings define how the model learns and determines an optimization algorithm, the loss function, and evaluation metrics.

In model training, Keras is used to train the NN model and each instruction in the above coding has the following meaning:

- **x_train**: Represents input data for training. In this case, it contains image data of handwritten digits from the MNIST dataset. The input data are entered into the input layer of the NN.
- **y_train**: Represents the correct answer data for training. In this case, it contains label data for handwritten digits from the MNIST dataset. The correct answer data are represented in a format that corresponds to the number of nodes in the output layer of the NN and are used when training the model.
- **epochs=5**: Specifies the number of model training operations. One epoch means that the model is trained with the entire training dataset for one time. In this case, the model is trained for five epochs.

The model.fit () method trains the model using the given training dataset. More specifically, the training data are split into minibatches and the gradient descent method is used for each minibatch to update model parameters. Then, the loss function and evaluation metrics for the model are calculated to report the progress of training. Splitting training data into minibatches means that a large dataset is split into several smaller batches to update model parameters for each minibatch. This allows the model to be trained efficiently without having to process all the data at once.

A power of two is usually specified as the number of samples for a minibatch. For example, 32, 64, and 128 are commonly used. Using minibatches can reduce the amount of memory required during the training process and parallelize computations. In many cases, a random selection of minibatches can help the model learn in a balanced manner.

The minibatch learning is often used when the stochastic gradient descent (SGD) method is used but can also be used with other optimization algorithms. Using minibatch learning enables efficient learning using a large dataset.

After five epochs, the model is expected to learn the entire training dataset and improve prediction performance. The output results are as follows:

```
< Start >
Downloading   data   from   https://storage.googleapis.com/tensorflow/tf-keras-
datasets/mnist.npz
11490434/11490434 [==============================]
        - 1s 0us/step
Epoch 1/5
1875/1875 [==============================] - 16s 8ms/step
        - loss: 2.5951 - accuracy: 0.8703
Epoch 2/5
1875/1875 [==============================] - 10s 5ms/step
        - loss: 0.3623 - accuracy: 0.9147
Epoch 3/5
1875/1875 [==============================] - 8s 4ms/step
        - loss: 0.2840 - accuracy: 0.9276
Epoch 4/5
1875/1875 [==============================] - 8s 4ms/step
        - loss: 0.2487 - accuracy: 0.9360
Epoch 5/5
1875/1875 [==============================] - 7s 4ms/step
        - loss: 0.2324 - accuracy: 0.9419
313/313 - 1s - loss: 0.2557 - accuracy: 0.9429 - 813ms/epoch
        - 3ms/step
Test accuracy: 0.9429000020027161
< End >
```

These messages indicate that TensorFlow is used to download data from the MNIST dataset and that the downloaded data are used to train the

NN model. The dataset contains images of handwritten digits and their corresponding labels (number classes).

This model was trained using five epochs. During one epoch, all the data in the training dataset are learned for one time. At the end of each epoch, the model returns a loss and a correct answer rate. The loss is a metric that measures the difference between model's predictions and actual labels. The correct answer rate indicates the percentage of data that the model predicted correctly.

Finally, it is indicated that the model has an accuracy of 0.9429, which means that the model can recognize handwritten digits with about 94.3% accuracy. This is a percentage at which the model can successfully classify digits accurately using the test dataset.

TensorFlow is an open-source machine learning library that provides low-level APIs for performing a variety of machine learning tasks. Keras is a high-level NN API for deep learning using TensorFlow as the back end. Keras is built on top of TensorFlow and hides TensorFlow's low-level operations, making it easier to build and train NNs.

1.3. Organization of this document

This book begins with the Basic, which describes data analysis methods using Excel. In the Basic, we will explain mean and variance testing, regression analysis using the Excel analysis tool, and how to implement an NN using the Excel solver. Furthermore, we will demonstrate automatic differentiation and introduce how to solve a simple PDE.

In the Practical Part that follows, we will explain how to create a spatiotemporal model using an NN. First, we will explain a method for visualizing page information from ancient literature data shot with a three-dimensional CT machine. Then, we will regard a magnetic line group calculated from the results of analyzing electromagnetic fields in a fusion reactor as physical data and explain how to visualize the plasma region using an NN. Up to this point, physical data irrelevant to time are handled.

As an advanced application, we will explain how to use an NN to derive a PDE that describes physical data and how to solve the PDE using the NN. Finally, we will demonstrate a surrogate model, together with some examples. The surrogate model streamlines what-if analysis by linking parameters, such as conditions from large-scale numerical simulations, with physical data.

Chapter 2

Basic

2.1. Background

2.1.1. *Data analysis using NNs*

This document describes the analysis of physical data primarily using NNs. An NN is a combination of mathematical models of neurons in the human cranial nervous system (Figure 2.1). The concept of NN is inspired by the mechanics of the human brain (all the interconnections between neurons). NN refers to a mathematical model created to represent some of the features of brain function on a computer. NN is one of the modeling methods in machine learning (ML) that mimics the workings of the human brain without the aim to model the billions of neurons in the brain accurately. It simply attempts to simplify it and make it operational. In NN, when one inputs data (numerical values), the numerical values are propagated with weights to the next layer. This is similar to the sequence of processes in the human brain, whereby synapses are connected with weights and neurons produce outputs, which are then connected to the next level.

Areas of use for NNs include pattern recognition and data mining (e.g., image recognition and recommendations).

NNs are widely used in video websites to analyze comments posted on videos. Specifically, NNs combine natural language processing techniques to perform sentiment analysis, topic classification, spam filtering, etc., of comments. The objective of sentiment analysis is to determine whether a comment falls into the positive, negative, or neutral category. NNs are useful for sentiment analysis because they can capture the meaning of words and sentences. Topic classification determines to which topic a comment relates. For example, to classify comments about a particular product or service, NNs can create a classifier using keywords related to that product

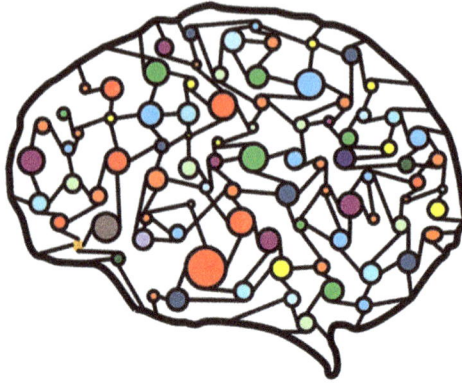

Figure 2.1. About NNs (https://ledge.ai/neural-network/).

or service. Spam filtering aims to detect automatically generated comments by spam and bots. By using NNs, characteristic patterns of spam comments can be detected and deleted automatically. Comment analysis using NNs can help video platforms control quality and improve products and services.

The introduction of NN-based methods in machine translation has dramatically improved translation accuracy. In machine translation, NNs are mainly used in a method called NMT (Neural Machine Translation). NMT uses an NN architecture called the Sequence-to-Sequence (Seq2Seq) model to encode input sentences into a latent semantic space, which is then decoded to produce output sentences.

Specifically, NMT uses two NNs called encoders and "decoders". The encoder receives input sentences as sequences of words or characters and converts them into a latent semantic space. The decoder receives as input the latent semantic space created by the encoder and decodes it to produce output sentences. In NMT, encoders and decoders use architectures such as RNN (Recurrent Neural Network), LSTM, and GRU. These architectures contribute to the accuracy of translation because they can handle long sequences and take into account the context needed for translation. NMT may also incorporate a mechanism called "attention" between the encoder and decoder. Attention is a mechanism that uses the output of the encoder to determine which part of the encoder's input sentence should correspond to which part of the decoder-generated sentence, and thus is useful when translating long or complex sentences.

In stock price prediction in the financial field, NNs are used to predict future stock prices based on past stock price data, economic indicators, and other information. In general, NNs use deep learning models such as MLP (Multilayer Perception) and LSTM (Long Short-Term Memory). There are two main methods of stock price forecasting. One is to forecast time series data and the other is to predict future trends, such as the rise or fall of a stock price.

The objective of forecasting time series data is to use past stock price data and other information to train NNs to predict future stock prices. For example, LSTM can be used to predict future stock prices by learning patterns of past stock price fluctuations.

Conversely, NN is used primarily as a classification problem in predicting trends such as rising or falling stock prices. The objective is to predict whether stock prices will rise or fall, based on past stock price fluctuations and other economic indicators. In this case, classifiers such as MLP (Multilayer Perception) and CNN (Convolutional Neural Network) are used.

However, since many factors affect stock price forecasts, it is difficult to make forecasts using only a single NN model. For this reason, ensemble learning, which combines multiple models to make forecasts, is often used.

Image recognition has become an indispensable technology in automated driving development. Image recognition systems using NNs are already in use to recognize surrounding objects faster and more accurately on behalf of the driver. In automated driving, NNs are used to recognize surrounding objects. A typical approach is to collect data from cameras, radar, lidar, and sensors and feed that information as input data to the NN. The NN then extracts feature values from the input data and uses them to recognize objects and extract information such as position, speed, and direction. In general, CNN (Convolutional Neural Network) is widely used for object detection and segmentation. Deep learning-based object detection models such as YOLO (You Only Look Once), SSD (Single Shot Detection), and Faster R-CNN are also used. These models can provide fast and accurate object detection. NNs are also used in other aspects of automation, such as vehicle tracking, lane recognition, and traffic sign recognition.

NNs can detect gastric cancer from endoscopic images, which allows them to be used in practice. NNs have achieved excellent results in image analysis and are used in various fields.

To detect gastric cancer, endoscopic images are input into NNs to identify regions where cancer is most likely to be present. Specifically, NNs can now automatically identify regions on a gastric endoscopic image where cancer is most likely to be present and detect gastric cancers 6 mm or larger with the same accuracy as a skilled endoscopist. In gastric cancer detection using NNs, it is important to train NNs using a large number of endoscopic images as training data. Once trained, the NNs can detect regions where cancer is likely to be present by inputting unknown endoscopic images.

NNs have also been applied to determine damage to paved roads, estimate the degree of damage inside bridges, detect abnormalities in power lines, etc. In paved road damage determination, data collected by vehicle vibration sensors and other devices are input into the NNs to estimate the state of pavement deterioration. In estimating the degree of damage inside bridges, data collected from vibration tests, sonic inspections, etc., are input into the NNs to assess the type and extent of the damage. In power line anomaly detection, data obtained from vibration sensors attached to the power lines are input into the NNs to detect abnormal vibrations. In this manner, NNs are used to detect various types of damage and abnormalities.

In agricultural work, harvesting and crop sorting heavily burden workers. Therefore, NNs are used in harvesting robots that assist farmers in the following ways:

Automatic Classification of Fruits and Vegetables: A harvesting robot automatically classifies harvested crops. NNs are used to analyze images of crops captured via a camera, to determine the type and ripeness of the crop.

Automated Harvesting: To harvest crops, a harvesting robot uses NNs to learn what ripe crops look like and then harvest them accurately. For example, a tomato harvesting robot learns the color and shape of ripe tomatoes, thereby accurately identifying and harvesting them.

Crop Health Monitoring: In a harvesting robot, NNs are used to monitor crop health. A harvesting robot uses a camera to capture images of the crop and uses NN to identify what diseases or insect damage the crop is infected with.

Crop Growth Prediction: NNs are used to predict crop growth. A harvesting robot uses data collected during crop growth to train the NNs and predict future growth. This allows farmers to predict the timing and quality of the harvest accurately.

As described, NNs play an important role in various fields. However, the training targets are mainly images and videos. NNs have rarely been used to describe physical data defined in spatiotemporal coordinates. PointNet, a set of spatial coordinate points, is a well-known NN for point cloud data (a set of points in 3D space).

PointNet is a type of NN that takes point cloud data as input data and extracts a global feature representation of the point cloud. PointNet was first introduced in 2017 by Qi CR, *et al.* [10] and is widely used in areas such as 3D object recognition, autonomous vehicles, and robot vision. A unique feature of PointNet is its ability to process point clouds directly. Conventional methods typically represent point clouds by converting them to voxel grids or 3D meshes and then applying a CNN. However, such methods can lead to compression and loss of information. PointNet avoids these problems by processing point clouds directly. PointNet can be used for tasks such as classification, segmentation, and object detection of point clouds. PointNet learns features such as point locations, colors, and normals from an input point cloud and combines these features to create a feature representation of the entire point cloud (Figure 2.2).

Using this global feature representation, PointNet can generate classification output data for a given point cloud data (Figure 2.3).

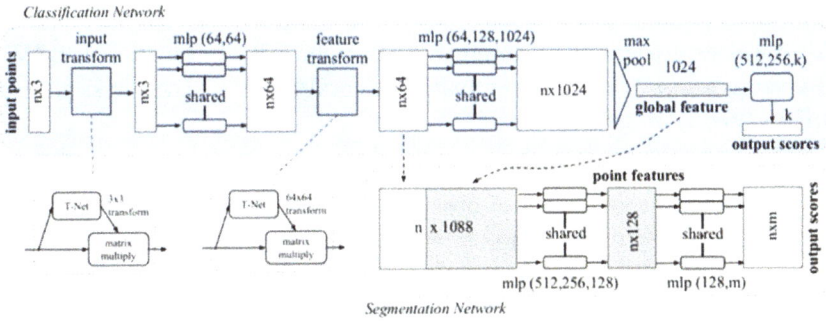

Figure 2.2. Overview of PointNet.

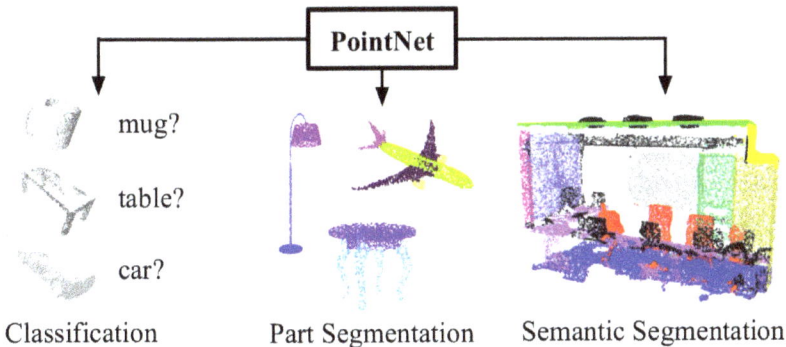

Classification   Part Segmentation   Semantic Segmentation

Figure 2.3. PointNet classification.

2.1.2. *Format of physical data*

Before describing the analysis and visualization of geophysical data, the format of geophysical data needs to be explained. Roughly stated, the format is the physical quantity defined on the point cloud data. In this document, two types of physical data are assumed: The first is a physical quantity u that is measured at a certain spatial location at a certain fixed time. A typical example is the data $u(t_i, x_j, y_j, z_j)$ obtained by a measurement device installed at a spatially fixed location (x_j, y_j, z_j) at time t_i. If there are N determined time(s) and M spatially fixed location(s), the physical data are denoted as follows:

$$u(t_i, x_j, y_j, z_j), \quad i = 1, N, \quad j = 1, M$$

u is scalar data defined by discrete spatiotemporal points but can also be vector or tensor data, depending on the instrument used for the measurement. When $N = 1$, it means discrete space data.

The second is data obtained at a specific spatiotemporal location. A typical example is data obtained by a mobile device (such as a UAV). If there is (are) N determined spatiotemporal point(s), the physical data are denoted as follows:

$$u(t_i, x_i, y_i, z_i), \quad i = 1, N$$

This N is called the number of observations.

When given physical data, interpolation or approximation is used to compute a spatiotemporal model that explains these data well. Interpolation creates a function that always passes through the given discrete data, and a typical interpolation method is kriging. If the given discrete data are likely to contain noise, it does not necessarily explain the original discrete data well. To solve such problems, approximations are sometimes used. Approximation does not require passing through the given discrete data and determines a function with as low an error as possible. A typical approximation method is regression. NN is a type of approximation and has more explanatory power than basic linear regression.

2.1.3. *Physical data visualization*

Physical data visualization refers to converting physical data into a form that is easily understood by humans through its representation in graphs and images. This allows us to visually understand the relationships and characteristics of the data, which is useful information for data analysis and research.

There are many ways to visualize physical data.

2.1.3.1. *1-D plot*

A 1-D plot is suitable when you are interested in only one parameter, such as time series data or statistical data. For example, plotting temperature changes along a time axis, with time on the x-axis and temperature values on the y-axis, provides a visual understanding of temperature changes. A 1-D plot includes line graphs, histograms, and box plots.

2.1.3.2. *2-D Plot*

A 2-D plot refers to plotting data using two axes (x-axis and y-axis). A 2-D plot is appropriate when you are interested in multiple parameters. For example, when examining the relationship between temperature and humidity, you can plot temperature on the x-axis and humidity on the y-axis to visually understand the relationship between temperature and humidity. A 2-D plot includes scatter plots, scatter plot matrixes, and contour plots.

2.1.3.3. *3-D Plot*

A 3-D plot refers to plotting data using three axes (x-, y-, and z-axes). A 3-D plot is suitable when you are interested in three or more parameters. For example, it is useful for displaying a bird's-eye view of temperature data acquired at the atmosphere's latitude, longitude, and height. Such data defined in three-dimensional coordinates are sometimes called volume data. Volume data refer to data that represent the position, shape, texture, etc., of a point or object in three-dimensional space. Volume data are used in a variety of fields, including medical imaging such as CT scans and MRIs, science and technology, architecture, design, and gaming. Volume data may be stored on a three-dimensional grid or expressed by a tetrahedral mesh. To visualize these data, expression methods such as 3D graphs and slice images are used. Also, when examining the relationship between temperature, humidity, and air pressure, you can plot temperature on the x-axis, humidity on the y-axis, and air pressure on the z-axis to visually understand the relationship. A 3-D plot includes isosurface plots and volume renderings.

Video: Animation visualization of continuous data along a time axis allows for a visual understanding of data changes.

Visualization Software: Visualization can be performed using programming languages such as Matplotlib, Python, and R. Specialized software such as Origin [11] and Grafana [12] can also be utilized.

VR/AR: VR/AR technology [13] can also be used to visualize data in three-dimensional space.

2.1.3.4. *Particle-based volume rendering*

For 3-D plots, this section describes one of the typical visualization methods, volume rendering [14]. This visualization method successfully represents the overall characteristics of the data, including its internal structure, by representing the target volume data as a semitransparent cloud. This section focuses on physical data visualization techniques based on volume rendering.

We will discuss PVBR (Particle-Based Volume Rendering) [15], in which the given physical data comprise opaque emitting particles to perform volume rendering. PBVR is a simple method that basically consists of two processing steps: particle generation and particle projection. Particle generation first requires an estimate of the particle density. PBVR was originally a visualization technique for continuously distributed physical data. Therefore, it was necessary to determine how to sample for that defined area. This time, however, the physical data are defined at discrete points. Thus, the particle density is determined by the originally given points. Next, the particle radius is determined by the user-specified transfer function for the opacity according to Equation (2.4).

The particle density ρ means the number of particles per unit volume. Consider a sphere of a certain radius and align each particle's position with the center of the sphere. The particles inside the sphere are counted and divided by the volume of the sphere to obtain the number density, which is the particle density ρ.

In ray casting, if the radius of the ray is r, consider a cylindrical interval of a certain length l along a ray and assume that the particles are Poisson distributed in this interval. The Poisson distribution is one of the probability distributions in probability theory and is used for random variables that take non-negative integer values, not continuous values. In particular, the Poisson distribution with mean λ is expressed by the following probability density function:

$$p(k; \lambda) = (\lambda^k e^{-\lambda})/k! \tag{2.1}$$

In this function, k represents a non-negative integer value $(0, 1, 2, 3 \ldots)$, λ represents the positive mean, e represents the base of the natural logarithm, and $k!$ represents the factorial of k.

The Poisson distribution is often used to represent the number of times independent events occur. For example, it is used to represent the number of times in which a certain event occurs in a given time frame, such as an accident or breakdown. The Poisson distribution is also used as an approximation for many events because the mean and variance are equal. In terms of physical data analysis and visualization, it is a means of expressing, for example, the probability of the number of particles generated in a particular spatial region.

Ray casting is a method of determining the pixels of an image along the line of sight from within a three-dimensional scene. This allows a three-dimensional scene to be projected onto a two-dimensional screen. Ray casting is often used to create realistic images because it allows physically accurate visualization of shadows, reflections, and other physical phenomena.

Light can travel along a ray without any obstruction in a space with a defined group of particles when the number of particles present in that space is zero. This probability is called transparency. The transparency t can be calculated as

$$t = e^{-\rho \pi r^2 l},\tag{2.2}$$

assuming that particles are generated according to the Poisson distribution described before since the volume is now $\pi r^2 l$ and the particle density is ρ. The sum of opacity and transparency is 1. Thus, the opacity can be calculated as follows:

$$\alpha = 1 - t = 1 - e^{-\rho \pi r^2 l}\tag{2.3}$$

The particle radius r can be calculated using the particle density ρ, the opacity value α, and the ray segment length Δt used in volume ray casting:

$$r = \sqrt{\frac{\log(1 - \alpha)}{\pi \rho \Delta t}}\tag{2.4}$$

The opacity value α is calculated by a user-specified transfer function. The transfer function shows the relationship between physical data values and opacity. A high opacity is set for data values that are to be emphasized, and a low opacity is set for physical data values that are not.

The first step in PBVR is to generate particles in the coordinates where the physical data are defined according to the particle radius given in the above equation. The second step is to project the generated particles onto the image plane. The particles are projected onto the image surface, and pixel values are calculated for each pixel. This basic processing step is repeated several times. The pixel values are added to the frame buffer and then divided by the number of iterations to obtain the final pixel values by averaging.

In PBVR, the projected images from the generated particles are added to the image plane to calculate the luminance values at the corresponding pixels. PBVR makes particles completely opaque so that when viewed by the eye, light, $c_i(i = 1, \ldots, n)$, from the particles at the back is blocked by the particles in the front. This effect can be achieved using the Z-buffer algorithm. The pre-projection sorting and alpha compositing processes typically required in volume rendering are unnecessary. The translucency effect is achieved in this averaging process.

In the projection process of the generated particles, the Z-buffer algorithm is used to keep the projected images of the particles closest to the eye position, and pixel values are calculated using these images. As particles are opaque, they do not require alpha compositing or reordering according to their distance from the eye position. For the last remaining particles, color mapping and shading calculations are performed. Color mapping uses a color transfer function to convert scalar data calculated by interpolation at the particle position into color data. The shading is determined by calculating the brightness based on the interpolated gradient vector and the light source vector, both of which are computed at the particle's position, and then multiplying this value by the color data. To calculate the final pixel values, we discuss ensemble averaging (Figure 2.4).

An ensemble is a set of results from multiple iterations of conceptually equivalent trial experiments. Each set of ensembles has the same attribute values for particle radius and particle density. In the ensemble averaging method, a trial experiment is assumed to be a single image creation process consisting of particle generation and projection with the same attribute values. The seed for random number generation is changed for each trial experiment. In ensemble averaging, the final pixel value is the average of the pixel values over all iterations. Therefore, if the i luminance value is B_i, the final pixel value is calculated as follows:

$$B^{\text{total}}(L_R) = \langle B^i \rangle = \sum_{i=1}^{L_R} \frac{B^i}{L_R} \tag{2.5}$$

In this equation, L_R represents the number of iterations and $B^{\text{total}}(L_R)^i$ represents the final luminance value from L_R iterations. Also, $\langle B^i \rangle$ represents the ensemble mean of B^i.

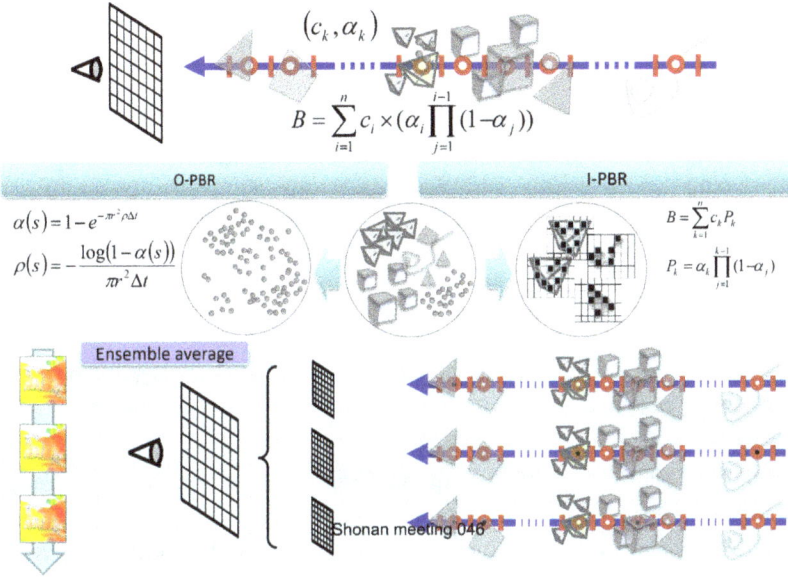

Figure 2.4. Pixel value calculation by ensemble averaging.

The variance of the final luminance value, with the luminance value B_i as the random variable, is expanded as follows:

$$B_{\text{Var}}^{\text{total}}(L_R) = \text{Var}\left(\sum_{i=1}^{L_R} \frac{B^i}{L_R}\right)$$

$$= \frac{1}{L_R^2}\left\{\sum_{i=1}^{L_R} \text{Var}\left(B^i\right) + 2\sum_{i,j,i<j} \text{Cov}\left(B^i, B^j\right)\right\} \quad (2.6)$$

The luminance values B_i are independent of each other. Thus, the value of the covariance $\text{Cov}(B_i, B_j)$ is 0. The variance of B_i can be determined as B_{var} regardless of the number of iterations. Thus, the standard deviation of the final luminance value is as follows:

$$B_{\text{Var}}^{\text{total}}(L_R) = \frac{B_{\text{Var}}}{L_R} \quad (2.7)$$

Note that the more the number of iterations used, the better the image quality becomes. This method is often used to improve the image quality of noisy images such as satellite images. The ensemble averaging method can be used to control the level of detail in the display of volume rendering results. Creating an average image with a small number of iterations is

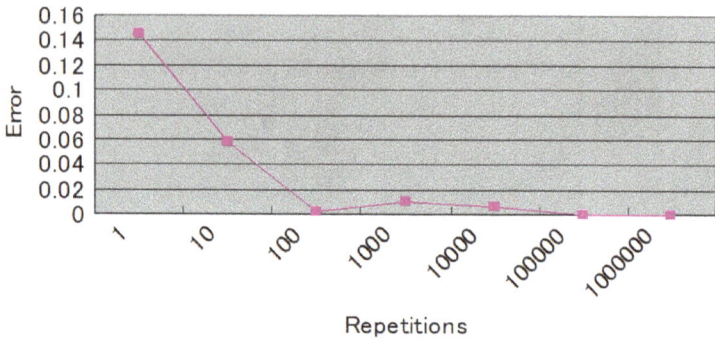

Figure 2.5. Relationship between the number of iterations and the quality of the generated images.

suitable for fast display. If high-quality rendering results are required, a sufficient number of iterations should be performed.

Figure 2.5 shows the relationship between the number of iterations and the quality of the generated image. You will note that the image quality improves as the number of iterations increases. In this case, image quality was evaluated by the average pixel value of the image difference from the ray-casted volume-rendered image created using the same transfer function.

2.2. Statistical Analysis in Excel

The following statistical analyses can be performed in Excel:

- Calculation of mean, median, maximum, minimum, standard deviation, variance, ratio, the ratio of change.
- Correlation analysis, regression analysis, t-test, F-test, ANOVA, Z-test, χ^2 test.
- Histograms, box plots, scatter plots, line charts, bar charts, pie charts, 3D charts.
- Pivot tables, contingency tables, conditional formatting, copying formulas, search and replace.
- Filtering, sorting, importing, and exporting data.
- Creating a macro, recording the operation history, saving and executing macros.

The following sections describe correlation analysis, regression analysis, t-test, F-test, and Z-test.

2.2.1. *Correlation analysis*

To perform a correlation analysis in Excel, insert the data and then select [Data Analysis] on the [Data] tab. Select [Correlation] in the dialog box to perform a correlation analysis. Correlation analysis allows you to calculate the relationship between two variables, which can then be expressed by a scatter plot or correlation coefficient. The correlation coefficient is a measure of how strong the relationship is between two types of data. It is expressed in the range of −1 to 1. The closer it is to 0, the weaker the correlation between variables, and the closer it is to 1, the stronger the correlation.

To display the correlation matrix in the Data Analysis tools in Excel, follow the given steps:

(1) Enter data into Excel.
(2) Click the [Data] tab and [Data Analysis].
(3) Select [Correlation].
(4) Select the entered data.
(5) Click [OK].

The correlation matrix will be displayed. The correlation matrix is a matrix of correlation coefficients between the data (Figure 2.6).

A correlation matrix is a square matrix of correlation coefficients between two or more variables. The correlation coefficient represents the strength of the linear relationship between variables and is expressed in the range of −1 to 1. Correlation matrices make it easy to visualize correlations between multiple variables. Correlation matrices are sometimes used to check correlations between variables using correlation matrices prior to regression analysis. If there are strong correlations among explanatory variables or between explanatory variables and the objective variable prior to

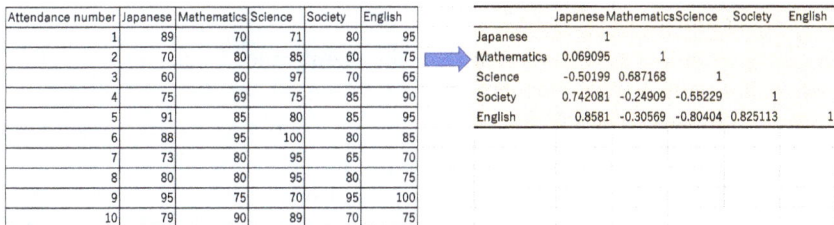

Attendance number	Japanese	Mathematics	Science	Society	English
1	89	70	71	80	95
2	70	80	85	60	75
3	60	80	97	70	65
4	75	69	75	85	90
5	91	85	80	85	95
6	88	95	100	80	85
7	73	80	95	65	70
8	80	80	95	80	75
9	95	75	70	95	100
10	79	90	89	70	75

	Japanese	Mathematics	Science	Society	English
Japanese	1				
Mathematics	0.069095	1			
Science	-0.50199	0.687168	1		
Society	0.742081	-0.24909	-0.55229	1	
English	0.8581	-0.30569	-0.80404	0.825113	1

Figure 2.6. Example of correlation matrix.

executing the regression analysis, multicollinearity problems may occur. Therefore, a correlation matrix is used to avoid that problem. When there are strongly correlated variables, the correlation matrix can be used to determine which of them to use.

2.2.2. *F-test*

In Excel, the F-test is a statistical test determining whether the difference in variances between two independent samples is significantly different. The F-test compares the ratio of the variances of two samples to determine if their means are significantly different. The F-test has higher detection power than the t-test when the two samples being compared have the same distribution. Therefore, it is often used when it is assumed that the two samples being compared have the same distribution.

When conducting a statistical test, formulate null and opposing hypotheses. The null hypothesis refers to the hypothesis that "the phenomenon does not exist", which is assumed when conducting a test. The opposing hypothesis refers to "the hypothesis that the phenomenon exists".

For example, when testing whether a drug is effective, the null hypothesis is set as "the drug is not effective" and the opposing hypothesis is "the drug is effective". The test results determine that the null hypothesis is rejected or supported.

In the F-test, a value called significance level is set. The significance level is a value that represents the "probability that the null hypothesis is true". Usually, the significance level is set to a value such as 0.05 or 0.01. For example, if the significance level is set at 0.05, and if the probability of the null hypothesis being true is less than 5%, the null hypothesis is rejected. In other words, the null hypothesis is judged to be "significantly different". Conversely, if the null hypothesis has a 5% or more probability of being true, the null hypothesis is supported. In other words, it is judged as "not significantly different".

The researcher themselves can set the significance level. However, 0.05 or 0.01 is usually used. The lower the significance level, the greater the statistical significance required to reject the null hypothesis. Conversely, the higher the significance level, the lower the statistical significance required to reject the null hypothesis.

No	A Group	B Group
1	31	29
2	29	31
3	30	32
4	32	30
5	28	
6	29	

F-Test: Two sample for Variances		
	A Group	B Group
Mean	29.83333333	30.5
Variance	2.166666667	1.666666667
Observations	6	4
df	5	3
F	1.3	
P(F<=f) one tail	0.441377641	
F Critical one-tail	9.013455168	

Figure 2.7. The results of the F-test.

To perform a two-sample F-test using the Data Analysis tools in Excel, follow the given steps:

(1) In Excel, open [File] and enable Analysis Tools.
(2) Select the [Data Analysis] tools from the [Data] tab.
(3) The "Data Analysis" dialog box appears. Select [F-test] and then click [OK].
(4) The "F-test" dialog box appears. Specify cell ranges for the data of the two populations in the [Variable 1 Range] and [Variable 2 Range]. Also, in "Output Range", specify the cell to which statics will be output.
(5) After clicking [OK], the results of the F-test are output to the specified cell (Figure 2.7).

The F-test is a statistical test method used to compare whether the variances of two or more populations are equal. The ratio of the variances of the two populations being compared is called the "variance ratio".

Specifically, in the F-test, independent samples from the two populations are obtained, and the variance of each sample is determined. The ratio of these variances is then calculated to obtain a value called the F-value. Based on whether this F value exceeds a certain boundary value, we determine whether the population variances are equal.

2.2.3. *t-Test*

The t-test, which can be performed in Excel, is a method for making comparisons when there is no correspondence between data elements (between two groups) or when there is correspondence between data elements (within one group). For comparisons between two groups, a prior F-test can be performed to determine whether the means of the two groups are significantly different by dividing them into groups where the variances can and cannot be considered equal. In a comparison within one group, you can examine whether the means of one group is significantly different from the other.

To perform a t-test in Excel, select [Data Analysis] from the [Data] tab, and then select [t-test: Paired Two-Sample for Means], [t-test: Two-Sample Assuming Equal Variances], or [t-test: Two-Sample Assuming Unequal Variances]. After specifying the data, the results of the t-test will be displayed.

To perform a two-sample t-test in the Data Analysis tools in Excel, follow the given steps:

(1) In Excel, open [File] and enable Analysis Tools.
(2) Select the [Data Analysis] tools from the [Data] tab.
(3) The "Data Analysis" dialog box appears. Select [t-test: Paired Two-Sample for Means], [t-test: Two-Sample Assuming Equal Variances], or [t-test: Two-Sample Assuming Unequal Variances], and then click [OK].
(4) The "t-test" dialog box appears. Specify cell ranges for the data of the two populations in the [Variable 1 Range] and [Variable 2 Range]. Also, in "Output Range", specify the cell to which statics will be output.
(5) After clicking [OK], the results of the t-Test are output to the specified cell (Figure 2.8).

Generally, a test of variance is performed prior to performing a test of the mean. If the resulting variances are not equal, use the [t-test:

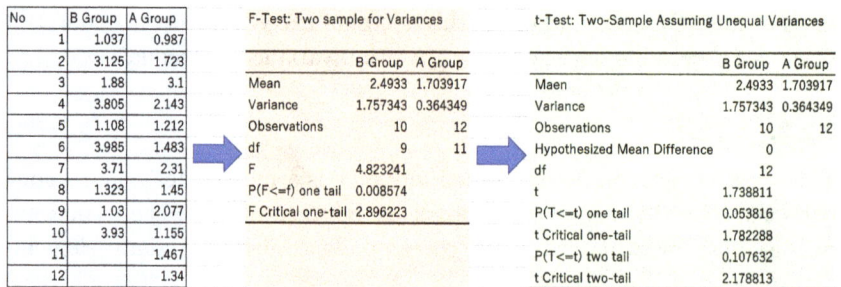

No	B Group	A Group
1	1.037	0.987
2	3.125	1.723
3	1.88	3.1
4	3.805	2.143
5	1.108	1.212
6	3.985	1.483
7	3.71	2.31
8	1.323	1.45
9	1.03	2.077
10	3.93	1.155
11		1.467
12		1.34

F-Test: Two sample for Variances

	B Group	A Group
Mean	2.4933	1.703917
Variance	1.757343	0.364349
Observations	10	12
df	9	11
F	4.823241	
P(F<=f) one tail	0.008574	
F Critical one-tail	2.896223	

t-Test: Two-Sample Assuming Unequal Variances

	B Group	A Group
Maen	2.4933	1.703917
Variance	1.757343	0.364349
Observations	10	12
Hypothesized Mean Difference	0	
df	12	
t	1.738811	
P(T<=t) one tail	0.053816	
t Critical one-tail	1.782288	
P(T<=t) two tail	0.107632	
t Critical two-tail	2.178813	

Figure 2.8. The Results of the t-Test.

Two-Sample Assuming Unequal Variances] to test the mean. In a t-test, the difference from the hypothetical mean represents the difference that exists between the population mean (or another sample mean) and the sample mean. Often, this value is set to zero. In t–testing, one-tailed tests and two-tailed tests are the testing methods used for hypothesis testing.

The one-tailed test is used to test whether the hypothetical mean is greater than or less than the sample mean. In other words, the one-tailed test determines whether the hypothesized mean is on the larger (upper) side or the smaller (lower) side. For example, if a drug is hypothesized to have an average 5-kg weight loss effect, a one-tailed test would determine whether the drug's efficacy is greater than 5 kg.

A two-tailed test, on the other hand, is used to test whether the hypothesized mean is different from the sample mean. In other words, a two-tailed test checks whether the hypothetical mean and the sample mean differ in either direction. For example, if a drug is hypothesized to have an average weight loss effect of 5 kg, a two-tailed test would determine whether the drug's efficacy differs from 5 kg.

One-tailed tests and two-tailed tests are used differently depending on the objective of the hypothesis test. One-tailed tests are used when testing a hypothesis that is restricted in one direction. The calculation of the p-value (the probability of observing extreme values above the test statistic obtained under the assumption that the null hypothesis is true) depends on whether the results are from an upper- or lower-tailed test. A two-tailed test, on the other hand, is used when the hypotheses may differ in either direction.

2.2.4. *Z-test*

In the Data Analysis tools in Excel, Z-test refers to a statistical test used to test the degree to which the sample mean differs from the population mean when the population mean is known. The Data Analysis tools in Excel provide tools to perform statistical tests, including Z-test. The Data Analysis tools in Excel allow you to test whether the population mean is significantly different from the sample mean.

2.3. Regression analysis

Regression analysis is a method for analyzing the degree to which a certain variable (the explanatory variable) x is affected by another variable (the explained variable) y. Regression analysis is often used for trend analysis and forecasting of data because it can search for statistically significant correlations. As we will discuss later, regression analysis can be used to derive a PDE. When the explanatory variable is a single variable, it is called simple regression, and when there are multiple explanatory variables, it is called multiple regression. We will discuss the simple regression model.

First, assume the following model:

$$y = \alpha + \beta x + u$$

In this equation, α and β represent certain constants, and u represents a random variable. That is, given x, the model assumes that y is determined by the deterministic term $\alpha + \beta x$ added to a random shock u. y, x, and u are referred to as follows:

- y: Explained variable, dependent variable.
- x: Explanatory variable, independent variable.
- u: Error term, disturbance term.

2.3.1. *Model characteristics*

The regression analysis model is a statistical method for modeling the relationship between an explanatory variable (independent variable) and an objective variable (dependent variable).

- The relationship between variables can be expressed by a linear model:

In regression analysis, the relationship between the explanatory variables and the objective variable can be expressed as a linear equation. The linear equation can be used to predict the value of the objective variable when given the explanatory variables. Although a linear model may seem restrictive, it is possible to capture the nonlinear effects of x by including x^2 as an explanatory variable or by performing variable conversion such as $\ln(x)$.

- The relationship between explanatory and objective variables can be elucidated:

Specifically, we can examine how the explanatory variables affect the objective variables and whether the effect is statistically significant.

- Allows for the assessment of how well the model fits:

In regression analysis, the degree to which a model fits the data can be evaluated. Specifically, we can evaluate how well the model fits the data by assessing the difference between the measured value and the value predicted from the model.

- The validity of the model can be verified:

In regression analysis, the validity of the model can be tested. Specifically, we can test whether the conditions assumed in the regression analysis are met.

- Multiple explanatory variables can be combined:

In regression analysis, multiple explanatory variables can be combined. By using multiple explanatory variables, complex relationships of effects on the objective variable can be modeled. A regression model with multiple explanatory variables is called multiple regression. A multiple regression model is a model that considers multiple factors simultaneously. The model is assumed to be as follows:

$$y = \alpha + \beta_1 x_1 + \beta_2 x_2 + \cdots + \beta_k x_k + u \tag{2.8}$$

2.3.2. *Regression analysis assumptions*

To estimate the value of (xy) based on the observed α and β, the following assumptions [16] are made:

(1) The relationship between the explanatory and objective variables is linear: Regression analysis assumes that the relationship between the explanatory and objective variables is linear. In other words, it is desirable that the data plotted on the scatterplot be linearly distributed; if not, a nonlinear regression model, etc., should be used.
(2) The error term should follow a normal distribution: Regression analysis assumes that the error in the forecast model follows a normal distribution. That is, the mean difference between the true and predicted values is assumed to be zero and follow a normal distribution with constant variance.

(3) The variance of the error term is equal for all i (homoscedasticity):

$$\text{var}\,(u_i) = \sigma^2 \tag{2.9}$$

Homoscedasticity refers to the fact that the variance of the error term is constant regardless of the value of the independent variable. In other words, it means that the scatter of the residuals plotted on a scatter plot is constant. If homoscedasticity is not met, the regression analysis results may be biased or skewed. For example, in regions of high variance, the regression line may deviate significantly from the actual values. Conversely, in regions where the variance is small, the regression line tends to be closer to the actual value.

(4) There is no multicollinearity among the explanatory variables: If there is a strong correlation between the explanatory variables, multicollinearity may occur. In such cases, the estimates of the regression coefficients may become unstable. To avoid multicollinearity, measures such as excluding highly correlated explanatory variables are necessary.

(5) Observed values are independent: Regression analysis assumes that the observed values are independent. That is, the impact of a certain observed value on the model should be independent of other observed values.

If these assumptions are not met, the results of the regression analysis may be biased or skewed. In order to obtain a more accurate predictive model, it is important to properly validate the assumptions before performing a regression analysis.

Assumption 1 is satisfied at the stage of formulating the linear equation.

To check Assumption 2, it is necessary to check whether some data (in this case, the error term) follow a normal distribution. One way to examine the normal distribution of data using Excel is to use a normal probability plot. The normal probability plot, described in the following, is a graph used to visually determine if the data are normally distributed.

2.3.2.1. *Visualization of normal probability plots*

There are two types of normal probability plots: normal *P-P* plot and normal *Q-Q* plot. Here, we will create a normal probability plot (normal *Q-Q* plot), as shown in Figure 2.9. The steps are from 1 to 8 (Reference: https://bellcurve.jp/statistics/blog/15362.html).

Normal Probability Plot

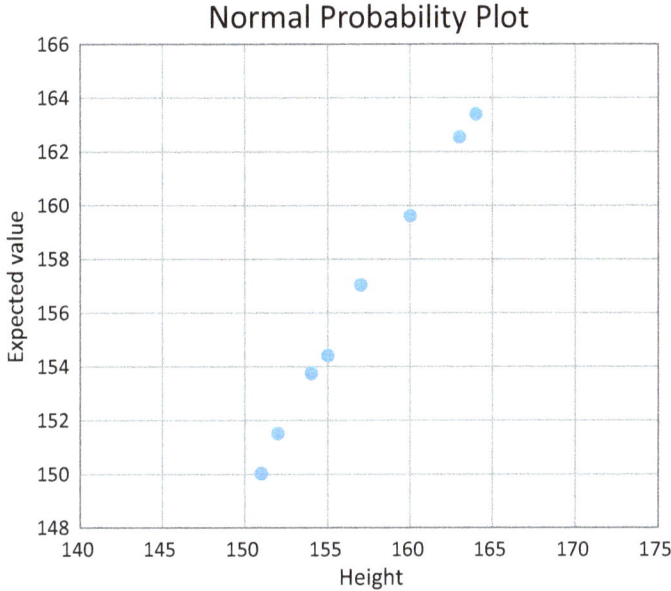

Figure 2.9. Normal probability plot.

Step 1: Measure the data.

The data on the left in Figure 2.10 are the results of height measurements of 200 female university students (*Source*: Toukei Kaiseki no Hanashi [Statistical Analysis Story]). We will use these data to create a normal probability plot.

Step 2: Rank the data.

Determine the rank of each piece of data using the RANK.EQ function.

$$=\text{RANK.EQ (cells, cell range, order)}$$

Specify the cell range as an absolute reference to the entire numerical range. If not specified, or if 0 is entered, the order is descending; if 1 is entered, the order is ascending. In this case, enter 1 (Figure 2.10).

Starting with the Excel 2010 version, two new functions have been added: RANK.EQ and RANK.AVG. The EQ and RANK.AVG functions return the highest and mean ranks, respectively, when the same number exists. In this case, we will use the RANK.EQ function.

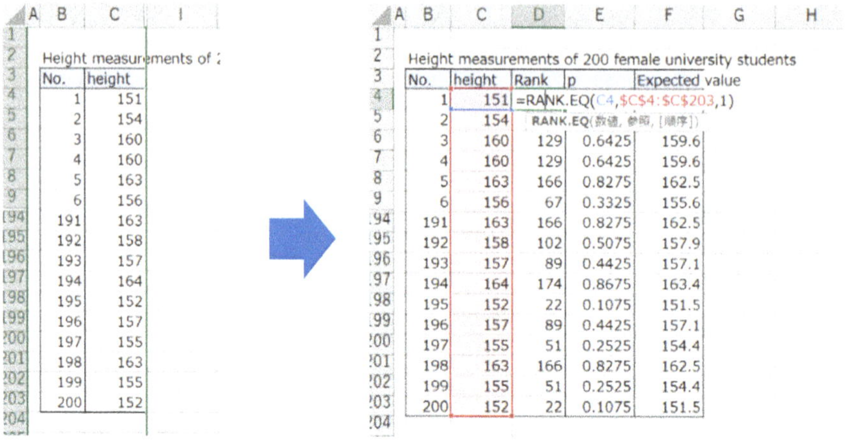

Figure 2.10. Data and their rankings.

Step 3: Convert the Rankings to Probability.
Convert the ranks to probability (Figure 2.10).

$$= (\text{rank} - 1/2)/n$$

This conversion formula has no logical basis and is one of several that have been proposed. However, since the determination of normality is based on visual inspection, there is no established theory as to which conversion formula is superior.

Step 4: Calculate the expected value from the probability.
Calculate the expected value of height using the NORM.INV function.

$$=\text{NORM.INV (probability,average,standard deviation)}$$

For the "argument probability", specify the probability obtained based on the rank obtained earlier. For "mean", specify the mean value of height, and for "standard deviation", specify the standard deviation of height (refer to Figure 2.11).

Step 5: Create a scatter plot.
Display the relationship between height and its expected value in a scatter plot (Figure 2.12).

Hold down the [Ctrl] key and select the cell range "C3:C203" (height) and "F3:F203" (z), then from the [Insert] tab, select [Charts] — [Insert Scatter (X, Y) or Bubble Chart] — [Scatter].

Figure 2.11. Probability of the data and their expected value.

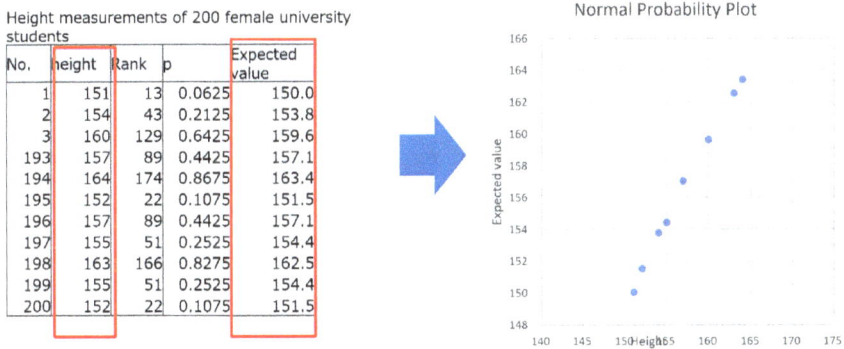

Figure 2.12. The data and their expected value.

Step 6: Enter a graph title.

Select [Graph Title] in the graph and enter a "graph title".

Step 7: Add a graph element.

When the graph is selected, the "Graph Tool" menu appears. Select the [Chart Design] tab, and then click [Add Chart Element] — [Axis Titles] — [Primary Vertical], and then enter the values in [Axis Title]. Repeat this process for the horizontal axis.

Store	Annual sales (10k yen)	Number of seats	Walking time from the station (分)	Breakfast (with=1, without=0)	Number of store-limited products	Residual
A	7800	54	6	0	2	-139.4182075
B	8718	75	3	1	4	-186.9401058
C	9401	80	1.5	1	5	113.8887097
D	8596	85	4	1	1	113.9421885
E	7235	40	7	0	0	-43.81207701
F	9396	76	1.5	1	6	7.804519739
G	7749	45	9	0	2	219.2921698
H	9288	70	2	1	6	44.3213885
I	7581	50	7	0	1	-36.06188526
J	8434	78	8	1	4	-93.01670057

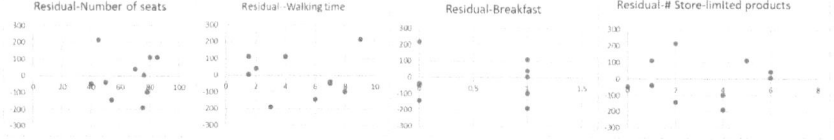

Figure 2.13. Scatter plots of explanatory variables and residuals.

Step 8: Complete the normal probability plot.

Complete the normal probability plot, including formatting as appropriate. If the points are aligned on a straight line $y = x$, the points can be considered as normally distributed.

To confirm Assumption 3, it is important to detect heterogeneity of variance. The following two methods are available for this detection:

- Check with scatter plots of explanatory variables and residuals (Figure 2.13).
- Check the scatter plot of the estimated value of the explained variable (= linear function of the explanatory variable) and the residuals.

The following Breusch–Pagan test has been widely proposed as a test for heterogeneity of variance.

2.3.2.2. *Breusch–Pagan test*

The Breusch–Pagan Test [17] is a method for testing whether the error terms in the residuals are independent. Specifically, the error terms are assumed to be independent in a regression analysis, but in reality, they may not be. The Breusch–Pagan Test can detect that the error terms are not independent in such cases. The resulting p-value indicates whether the error terms are independent or not.

Assume now that the multiple regression model is formulated as follows:

$$y_i = \alpha + \beta_1 x_{1,i} + \beta_2 x_{2,i} + \cdots + \beta_k x_{k,i} + u_i \qquad (2.10)$$

The error is determined as follows:

$$e_i = y_i - \alpha - \beta_1 x_{1,i} - \beta_2 x_{2,i} - \cdots - \beta_k x_{k,i} \qquad (2.11)$$

Assume that the square error e_i^2 is formulated by a similar regression model:

$$e_i^2 \quad = \quad \delta_0 + \delta_1 x_{1,i} + \delta_2 x_{2,i} + \cdots + \delta_k x_{k,i} + v_i \qquad (2.12)$$

Then, formulate the following null hypothesis H_0:

$$H_0: \ \delta_1 = \delta_2 = \cdots = \delta_k = 0 \qquad (2.13)$$

The test for H_0 is performed by an F-test using the following F-value:

$$\frac{(\text{RSS} - \text{TSS})/k}{\text{RSS}/(n - (k+1))} = \frac{(\text{ESS})/k}{\text{RSS}/(n - (k+1))} \sim F(k, n - (k+1)) \qquad (2.14)$$

The resulting p-value indicates the probability that the hypothesis is true if the null hypothesis is that "the variance of the residuals is independent of the explanatory variable (that is, satisfies the homoscedasticity)". The smaller the p-value, the more likely it is that the null hypothesis is rejected, and the variance of the residuals depends on the explanatory variable (that is, it does not satisfy the homoscedasticity).

If the p-value is somewhat small (usually less than 0.05 or less), the assumption of homoscedasticity is not satisfied. Therefore, it is necessary to consider improving the model or employing a different statistical method. On the other hand, if the p-value is sufficiently large, the assumption of homoscedasticity is considered to be satisfied, and the results of the regression analysis can be used with confidence. When we perform a regression analysis using the square of the residuals as the objective function for the sales data described in Figure 1.12, we see that the F-value is 0.9472, which is sufficiently high (Figure 2.14).

2.3.3. *Regression analysis using the data analysis tools in Excel*

To perform a regression analysis using the Data Analytical tools in Excel, follow the given steps:

(1) Enter the data into an Excel sheet.
(2) Select the data, and click [Data Analysis] from the [Data] tab.
(3) In the "Data Analysis" dialog, click [Regression]. In the "Regression" dialog, select the explanatory variables in the [Input X Range] and the objective variables in the [Input Y Range]. Select a regression model and set options as needed.

Store	Annual sales (10k yen)	Number of seats	Walking time from the station (Min)	Breakfast (with =1, without =0)	Number of store-limited products	Residual	Residual^2
A	7800	54	6	0	2	-139.4182075	19437.43658
B	8718	75	3	1	4	-186.9401058	34946.60317
C	9401	80	1.5	1	5	113.8887097	12970.6382
D	8596	85	4	1	1	113.9421885	12982.82232
E	7235	40	7	0	0	-43.81207701	1919.498092
F	9396	76	1.5	1	6	7.804519739	60.91052835
G	7749	45	9	0	2	219.2921698	48089.05572
H	9288	70	2	1	6	44.3213885	1964.385478
I	7581	50	7	0	1	-36.06188526	1300.459568
J	8434	78	8	1	4	-93.01670057	8652.106586

概要

回帰統計	
重相関 R	0.341570915
重決定 R2	0.11667069
補正 R2	-0.589992758
標準誤差	20187.94442
観測数	10

分散分析表

	自由度	変動	分散	観測された分散比	有意 F
Regression	4	269149348	67287337	0.1651	0.9472
Residual	5	2.038E+09	407553100		
Summation	9	2.307E+09			

Figure 2.14. Breusch–Pagan test.

(4) In Output Range under Output Options, select the cell in the left corner of the area where the results will be output.

(5) Click [OK] to output the regression analysis results (Figure 2.15).

Regression analysis can be easily performed by using the Data Analysis tools in Excel. From the regression analysis results, you will note the impact of the explanatory variables on the objective variable. The Data Analysis tools in Excel can also be used to evaluate the analysis results with a regression model. The evaluation measures include the followings:

• Correction "R^2" to check the accuracy of the estimated regression equation: In regression analysis, a measure called the coefficient of determination is used. The coefficient of determination is a measure of

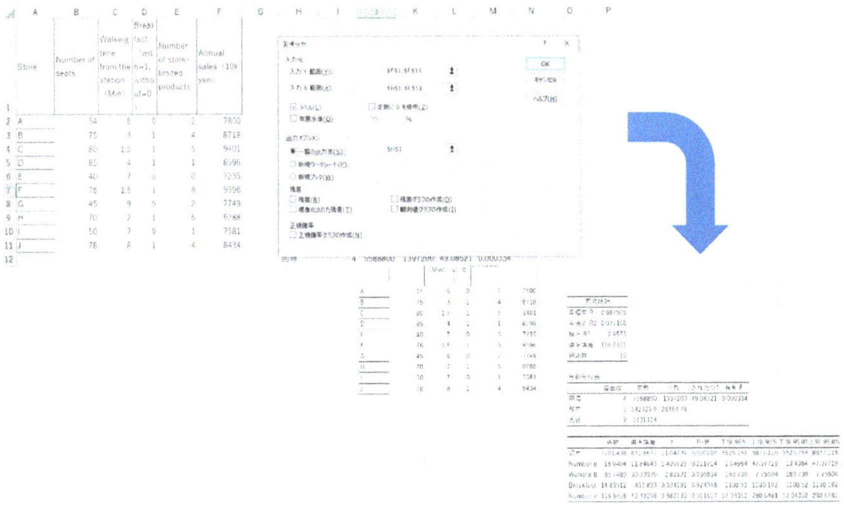

Figure 2.15. Regression analysis example.

how well the regression model fits the data and takes values between 0 and 1. A value closer to 1 means that the regression model fits the data well, while a value closer to 0 means that the regression model does not fit the data well. The following equation calculates the coefficient of determination: $R^2 = 1 - (\text{SSE}/\text{SST})$. In this equation, SSE represents the residual sum of squares (the sum of squares of the difference between the value predicted by the regression model and the observed value) and SST represents the sum of squares of hypotheses (sum of squares of the difference between the observed value and the mean value). The coefficient of determination is often used to evaluate the results of regression analysis. However, the coefficient of determination tends to increase in value as the number of explanatory variables is increased. Therefore, the complexity of the model must be taken into account.

- Check if the estimated regression equation is statistically meaningful (Significant F): Significance F is a measure used in regression analysis to determine whether the regression model fits the data better. Significance F is obtained by calculating the ratio of the residual sum of squares (SSE) to the sum of squares of hypotheses (SST). SSE is the sum of squares of the difference between the value predicted by the regression model and the observed value, and SST is the sum of squares of the difference between the observed value and the mean value. Significance F is greater when SSE is small (that is, the regression model fits the data well) and SST is large (that is, the observed values deviate significantly from the mean). To calculate the significance F, use the following equation:

$$F = (\text{SST} - \text{SSE})/(\text{SSE}/(n - p - 1)) \qquad (2.15)$$

In this equation, n represents the sample size, and p represents the number of explanatory variables. Significance F calculates the probability that the data follow an F distribution and compares it to a significance level (usually 0.05, 0.01, etc.) to determine if the regression model significantly fits the data. If the significance F is less than the significance level, the regression model is significantly fitting the data.

- Check if the estimated coefficient is statistically meaningful (p-value): The larger the absolute value of the estimated coefficient from 0, the easier it is to see that it is statistically meaningful. If the p-value is small (usually less than 0.05), the probability that the coefficient is 0 is small, meaning that the null hypothesis that the coefficient is 0 is rejected.
- Check the degree of influence of each explanatory variable (coefficient): In regression analysis, the following methods are used to select the explanatory variables to be used for the analysis:

  - Select explanatory variables that are considered significant based on prior knowledge and theoretical grounds.
  - Check the correlation coefficients among the explanatory variables. Explanatory variables with high correlations can be eliminated because they can be confusing in interpreting regression models.
  - Using univariate regression analysis, check the correlation coefficient of each explanatory variable with the objective variable. Select explanatory variables with high correlation coefficients.
  - Automatically select explanatory variables using measures such as multicollinearity condition index ratio and significance F. The multicollinearity condition index is a measure that refers to the state of high correlation between explanatory variables in multiple regression analysis.
  - When interpreting the results of a regression analysis, you can also assess how well the regression model fits the data by checking the p-values and coefficients of determination of the explanatory variables.

2.3.3.1. *Variable reduction method*

Variable reduction methods [18], often used in regression analysis, use p-values to select meaningful explanatory variables. In regression analysis, a method called variable reduction allows us to simplify the model by removing unnecessary variables. This not only makes the model easier to interpret but also prevents overlearning. It can also be used to derive a PDE from physical data, which will be discussed later. A PDE can be derived by searching for a model in which a partial differential term (often a time partial differential term) is explained by several other differential terms (often spatial partial differential terms). Using variable reduction methods, it is possible to delete unnecessary partial differential terms from many candidates.

Variable reduction with p-values is a method that uses statistical methods to select explanatory variables used in regression analysis that are significantly related to the explained variable. The p-value represents the probability that a hypothesis is true. p-value-based variable reduction methods determine whether each explanatory variable is significantly related to the explained variable based on the p-value of the regression model using that explanatory variable. If the p-value is smaller than the significance level (usually 0.05 or 0.01), the explanatory variable is interpreted as significantly related to the explained variable. Therefore, if the p-value is above the significance level, the explanatory variable is deleted from the model.

The variable reduction method using p-values is a type of backward selection method. It has the advantage of being less sensitive to the order in which variables are added since all explanatory variables can be evaluated simultaneously, rather than adding variables one by one. Using the Data Analysis tools in Excel, regression analysis can be converted to a macro, and the steps described earlier can be performed in VBA to create an automatic variable reduction method.

2.3.3.2. *Autoregression*

The Data Analysis tools in Excel allow you to perform autoregressions [19]. An AR (autoregressive) model is one of the methods to analyze time series data. Autoregressive models allow you to analyze trends and periodicity of time series data by assuming that the current values of time series data are determined by past data values and by building such models. The autoregressive model is expressed by the following equation:

$$y(t) = c + \varphi 1 y(t-1) + \varphi 2 y(t-2) + \cdots + \varepsilon(t) \qquad (2.16)$$

In this equation, $y(t)$ represents the current value of the time series data, c represents a constant, $\phi 1, \phi 2, \ldots$ represent the autoregressive coefficients, and $\varepsilon(t)$ represents the error term. The larger the autoregressive coefficient, the stronger the influence of the past data values. The error term represents the random noise in the time series data. Autoregressive models may be used for predicting or filtering time series data.

To perform an autoregression using Regression in the Data Analysis tools in Excel, follow the given steps: Use actual data recorded hourly of the water level of the Oyodo River at the Hiwatashi point on September 16, 1990.

(1) Insert time data in the first column and water level data in the second column. Time is equally spaced from top to bottom (Figure 2.16).

Time	Water level
9/16/90 9:00	0.29
9/16/90 10:00	0.27
9/16/90 11:00	0.26
9/16/90 12:00	0.26
9/16/90 13:00	0.26
9/16/90 14:00	0.26
9/16/90 15:00	0.27
9/16/90 16:00	0.27
9/16/90 17:00	0.27
9/16/90 18:00	0.27
9/16/90 19:00	0.27
9/16/90 20:00	0.27
9/16/90 21:00	0.29
9/16/90 22:00	0.3

Figure 2.16. River water level data.

Time	Water level	1H before	2H before	3H before	4H before	5H before	6H before	7H before	8H before	9H before	10H before
9/16/90 9:00	0.29										
9/16/90 10:00	0.27	0.29									
9/16/90 11:00	0.26	0.27	0.29								
9/16/90 12:00	0.26	0.26	0.27	0.29							
9/16/90 13:00	0.26	0.26	0.26	0.27	0.29						
9/16/90 14:00	0.26	0.26	0.26	0.26	0.27	0.29					
9/16/90 15:00	0.27	0.26	0.26	0.26	0.26	0.27	0.29				
9/16/90 16:00	0.27	0.27	0.26	0.26	0.26	0.26	0.27	0.29			
9/16/90 17:00	0.27	0.27	0.27	0.26	0.26	0.26	0.26	0.27	0.29		
9/16/90 18:00	0.27	0.27	0.27	0.27	0.26	0.26	0.26	0.26	0.27	0.29	
9/16/90 19:00	0.27	0.27	0.27	0.27	0.27	0.26	0.26	0.26	0.26	0.27	0.29
9/16/90 20:00	0.27	0.27	0.27	0.27	0.27	0.27	0.26	0.26	0.26	0.26	0.27
9/16/90 21:00	0.29	0.27	0.27	0.27	0.27	0.27	0.27	0.26	0.26	0.26	0.26
9/16/90 22:00	0.3	0.29	0.27	0.27	0.27	0.27	0.27	0.27	0.26	0.26	0.26
9/16/90 23:00	0.3	0.3	0.29	0.27	0.27	0.27	0.27	0.27	0.27	0.26	0.26
9/17/90 0:00	0.29	0.3	0.3	0.29	0.27	0.27	0.27	0.27	0.27	0.27	0.26
9/17/90 1:00	0.29	0.29	0.3	0.3	0.29	0.27	0.27	0.27	0.27	0.27	0.27
9/17/90 2:00	0.29	0.29	0.29	0.3	0.3	0.29	0.27	0.27	0.27	0.27	0.27
9/17/90 3:00	0.29	0.29	0.29	0.29	0.3	0.3	0.29	0.27	0.27	0.27	0.27
9/17/90 4:00	0.31	0.29	0.29	0.29	0.29	0.3	0.3	0.29	0.27	0.27	0.27

Figure 2.17. Copy of historical water level data.

(2) Copy all data in Column 2 from Row 2 and subsequent rows, and paste them into Column 3, one row down from Row 2. Repeat this step for the required time interval. Assume that this step is repeated N times. The second column holds the water level, which is the objective variable, and the subsequent columns hold the explanatory variables (historical water level data) (Figure 2.17).

(3) At this stage, Regression under the "Data Analysis" tools is performed. For the objective variable, select from the cell moved N rows down. Assuming $N = 5$, create a model to predict water levels using water level data from 1 hour to 5 hours ago. The original and predicted data are listed in the 2nd and 13th Columns, respectively. Autoregression was successfully performed by regression analysis (Figure 2.18).

2.3.4. *Excel macro*

An Excel macro is a program used in Excel. In Excel, macros can be created using a programming language called VBA (Visual Basic for Applications).

Macros can be used to automate repetitive tasks and operations in Excel, which is very useful when processing large amounts of data. Macros can also be used to extend the functionality of Excel.

Time	Water level	1H before	2H before	3H before	4H before	5H before	6H before	7H before	8H before	9H before	10H befor	Prediction
9/16/90 9:00	0.29											
9/16/90 10:00	0.27	0.29										
9/16/90 11:00	0.26	0.27	0.29									
9/16/90 12:00	0.26	0.26	0.27	0.29								
9/16/90 13:00	0.26	0.26	0.26	0.27	0.29							
9/16/90 14:00	0.26	0.26	0.26	0.26	0.27	0.29						
9/16/90 15:00	0.27	0.26	0.26	0.26	0.26	0.27	0.29					
9/16/90 16:00	0.27	0.27	0.26	0.26	0.26	0.26	0.27	0.29				
9/16/90 17:00	0.27	0.27	0.27	0.26	0.26	0.26	0.26	0.27	0.29			
9/16/90 18:00	0.27	0.27	0.27	0.27	0.26	0.26	0.26	0.26	0.27	0.29		
9/16/90 19:00	0.27	0.27	0.27	0.27	0.27	0.26	0.26	0.26	0.26	0.27	0.29	
9/16/90 20:00	0.27	0.27	0.27	0.27	0.27	0.26	0.26	0.26	0.26	0.27		0.27
9/16/90 21:00	0.28	0.27	0.27	0.27	0.27	0.27	0.26	0.26	0.26	0.26		0.26
9/16/90 22:00	0.3	0.28	0.27	0.27	0.27	0.27	0.27	0.26	0.26	0.26		0.26
9/16/90 23:00	0.3	0.3	0.28	0.27	0.27	0.27	0.27	0.27	0.26	0.26		0.26
9/17/90 0:00	0.29	0.3	0.3	0.28	0.27	0.27	0.27	0.27	0.27	0.26		0.26
9/17/90 1:00	0.29	0.29	0.3	0.3	0.29	0.27	0.27	0.27	0.27	0.27		0.27
9/17/90 2:00	0.29	0.29	0.29	0.3	0.3	0.29	0.27	0.27	0.27	0.27		0.27
9/17/90 3:00	0.29	0.29	0.29	0.29	0.3	0.3	0.29	0.27	0.27	0.27		0.27
9/17/90 4:00	0.31	0.29	0.29	0.29	0.29	0.3	0.3	0.29	0.27	0.27		0.27
9/17/90 5:00	0.32	0.31	0.29	0.29	0.29	0.29	0.3	0.3	0.29	0.27		0.27
9/17/90 6:00	0.33	0.32	0.31	0.29	0.29	0.29	0.29	0.3	0.3	0.29		0.27
9/17/90 7:00	0.33	0.33	0.32	0.31	0.29	0.29	0.29	0.29	0.3	0.3		0.29

概要

回帰統計

重相関 R	0.999672
重決定 R2	0.999145
補正 R2	0.999106
標準誤差	0.057762
観測数	116

分散分析表

	自由度	変動	分散	観された分散	有意 F
回帰	5	428.8593	85.77186	25707.21	5.8E-167
残差	110	0.367014	0.003336		
合計	115	429.2263			

	係数	標準誤差	t	P-値	下限 95%	上限 95%	下限 95.0%	上限 95.0%
切片	0.011023	0.00777	1.41867	0.158822	-0.00438	0.026422	-0.00438	0.026422
X 値 1	2.119435	0.091103	23.26422	2.76E-44	1.938891	2.29998	1.938891	2.29998
X 値 2	-1.28363	0.21589	-5.94575	3.31E-08	-1.71147	-0.85578	-1.71147	-0.85578
X 値 3	0.392878	0.245412	1.600895	0.112268	-0.09347	0.879226	-0.09347	0.879226
X 値 4	-0.52781	0.215848	-2.4453	0.01606	-0.95557	-0.10005	-0.95557	-0.10005
X 値 5	0.293442	0.090918	3.227528	0.001646	0.113263	0.473621	0.113263	0.473621

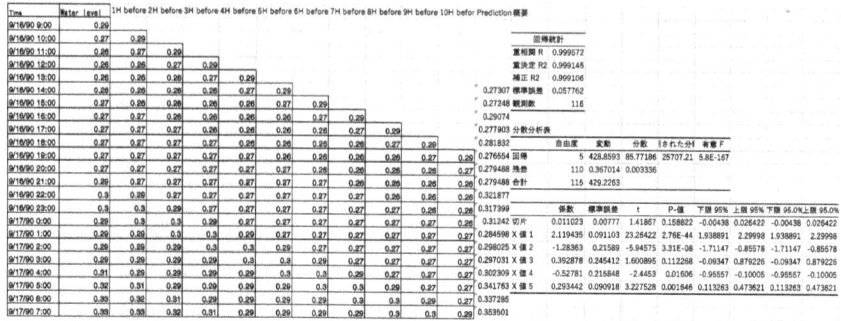

Figure 2.18. Water level prediction results from regression analysis.

To create macros, you will need to learn VBA. To learn VBA, basic knowledge of programming is helpful. However, even those who have never studied programming will be able to create macros by learning step by step.

To use macros in Excel, you must enable them in the settings. The steps are as follows:

(1) Open the [Developer] from the menu and click [Macro Security].
(2) Click [Macro Settings].
(3) In the Macro Settings, select [Enable VBA macros].
(4) Click the [OK] button to save the settings.

You will now be able to use macros in Excel.

Note: Enabling macros may result in the risk of dangerous macros running on your computer. Before enabling, be sure to verify that the macros you are running are from trusted sources.

The "Record Macro" button may not appear in Excel immediately after installation. If this is the case, to enable macros in Excel, follow the given steps:

(1) Open the [File] menu and click [Options].
(2) Click the [Customize Ribbon] from the menu.
(3) In the "Popular Commands" list, select [Developer] and press the [OK] button.
(4) Click the [OK] button to save the settings (Figure 2.19).

To record a macro in Excel, follow the given steps:

(1) Click the [Developer] tab.
(2) Click the [Record Macro] button.

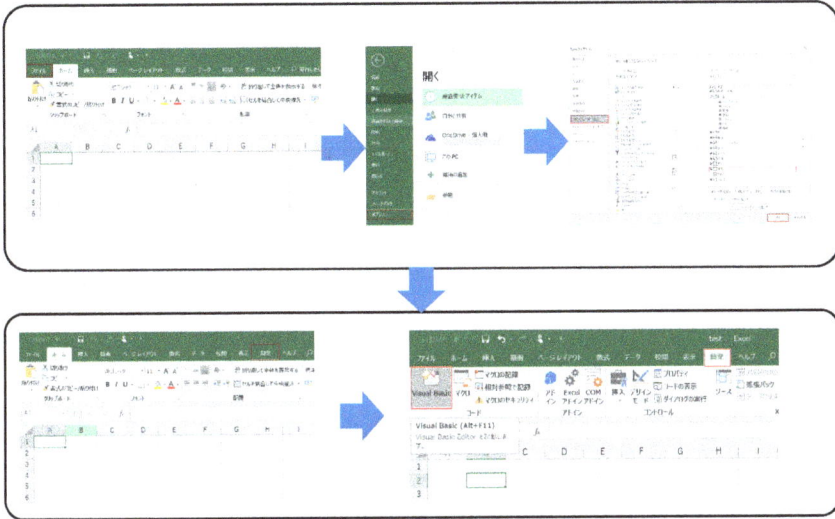

Figure 2.19. How to enable macros.

(3) Recording starts. Your mouse and keyboard operations are now being recorded. Perform your tasks in Excel.

(4) To end the recording, click the [Record Macro] button again.

(5) The "Record Macro" dialog box appears. Enter a name for the recorded macro and click the [OK] button.

You have now recorded a macro in Excel.

Note: When you record a macro in Excel, the VBA code is automatically generated to reproduce the recorded operation. To run a recorded macro, go to the [Developer] tab, click the [Macros] button, and select the recorded macro, or click the "Run Macro" button.

VBA (Visual Basic for Applications) is a programming language used in Microsoft Office applications (Excel, Word, Outlook, PowerPoint, etc.). VBA can be used to automate repetitive tasks and operations in Excel. VBA can also be used to extend the functionality of Excel. VBA is automatically generated when you record a macro in Excel. However, you can also write VBA directly. By writing VBA directly, you can create more complex, customized macros. To learn VBA, basic knowledge of programming is helpful. However, even those who have never studied programming will be able to create VBA by learning step by step.

VBA uses a similar syntax to Visual Basic. The basic syntax of VBA includes the following:

Variables: Stores data. It can store a variety of data types, including numbers and strings.

Control Syntax: Syntax for controlling a program, including If and For statements.

Function: Performs a specific operation. VBA provides standard functions (e.g., SQR and LEFT functions). You can also create your own functions.

Statements: A single statement. VBA requires a semicolon (;) at the end of a statement.

These are the basics of VBA syntax. In addition, VBA allows you to use methods and properties to take advantage of various Excel functions.

For example, a VBA program that calculates the maximum value and its position among multiple data would look like this:

```
<Begin>
Sub FindMax()
    Dim data() As Variant 'Array to store the data
    Dim i As Long 'Counter variable
    Dim maxValue As Variant 'Maximum value
    Dim maxIndex As Long 'Position of the maximum value
    'Store the data in the array
    data = Array(10, 20, 30, 40, 50)
    'Find the maximum value and its position
    maxValue = data(0) 'Initialize the maximum value with the
        first element
    maxIndex = 0 'Initialize the position of the maximum value with the
        first element
    For i = 1 To UBound(data)
        If maxValue < data(i) Then
            maxValue = data(i) 'Update the maximum value
            maxIndex = i 'Update the position of the maximum value
        End If
    Next i
    'Display the results
    The "maximum value" in the MsgBox is "& maxValue &", and its
        position is "& maxIndex &".
End Sub
<End>
```

In the above program, data are stored in the array data. One can then use a For statement to examine the contents of the array one by one. If the current element is greater than the maximum value, the maximum value is updated, and its position is also updated.

Finally, the results are displayed using the MsgBox function, which prints the maximum value and its position. To execute the above program, write the above-mentioned code in the VBA editor and press the [F5] key or click the [Run Macro] button.

This program can be used to find the largest p-value of an explanatory variable resulting from a regression analysis that was output on an Excel sheet by deleting that explanatory variable and then executing the regression analysis again.

2.3.5. *Implementation of the variable reduction method using Excel VBA*

The variable reduction method described in the regression analysis is implemented using VBA. In this implementation, the process is to find the smallest absolute value of the *t*-value in place of the *p*-value and repeat the regression analysis while removing that variable. At the 5% level of the *p*-value, the absolute value of the *t*-values will generally be 2 or more and negatively correlated with each other.

Assume that the Excel sheet holds data for the objective variable in the first column and data for the explanatory variable in the second and subsequent columns for the number of explanatory variables. Assume the first row has the number of explanatory variables entered in the second column and the number of sample sizes in the first column. The results of the regression analysis (coefficients for the explanatory variables, *p*-values, and *t*-values) should be output with the top left cell (labeled Summary) spaced one row apart from the last data row.

The following is an example of applying the variable reduction method to the survey results to narrow down the objective variables that significantly affect the objective function. In this survey, 106 first-year university students were asked to respond to an interval scale on nine different items related to their level of satisfaction and expectations of the university. An interval scale is a quantitative measurement scale that is not ordinal, the difference between two variables is meaningful and equal, and the presence of zero is optional. When you execute the VBA program for the variable reduction method, you find that the two essential items remain (Figure 2.20).

Analysis and Visualization of Discrete Data

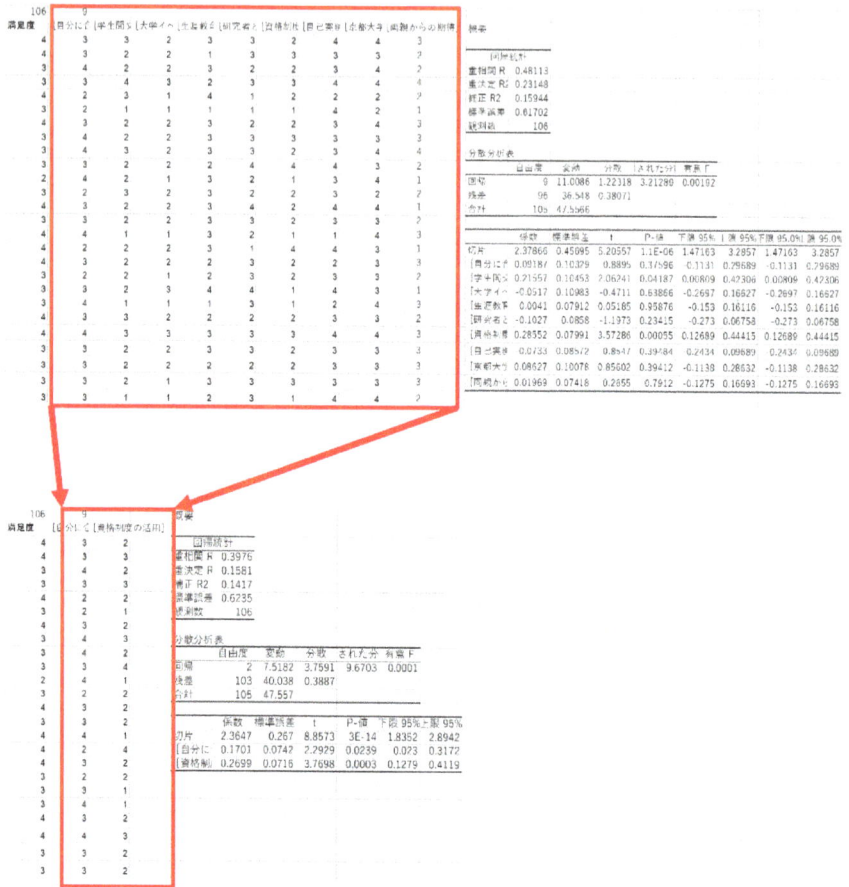

Figure 2.20. Example of the analysis of survey results using variable reduction method.

To prepare, first obtain the macro output that executes Regression under the Data Analysis tools, deletes columns, and left-aligns the text. Turn on Record Macro and execute the Regression under the Data Analysis tools, then delete the columns, and execute Align Left. As you will note from the results, Excel-specific settings for specifying rows and columns are required. To implement this in a VBA program, you need to learn how to manipulate character-type data. We will now discuss how to do this.

There are four types of character-type data manipulation required in this program implementation.

- Obtain character codes from character-type data: To assign the character code of character-type data A to the third column of the first row, write the following: Cells(1,3) = Asc("A")
- Convert from character code to character-type data: To assign character-type data with character code 66 to the fourth column of the first row, write the following: Cells(1,4)=Chr(66)
- Convert integer-type data to character-type data: To convert integer-type data NROWS + 4 to character-type data and assign this to the fifth column of the first row, write the following: Cells(1,5)= str(NROWS + 4)
- Concatenate strings: To concatenate and assign three character-type data NR in the sixth column of the first row, write the following: Cells(1,6)="B2:"+"L"+"20"

To search for the explanatory variable data column with the smallest absolute value of t-value, write the following in the VBA program.

```
<Begin>
TMIN = 999999
   For j = 1 To CVARS
       TVAL = Abs(Cells(j + NLINES + 20, 4))
       If TVAL < TMIN Then
          TMIN = TVAL
          IMIN = j
       End If
   Next j
<End>
```

CVAR represents the number of explanatory variables, NLINES represents the sample size, and TMIN represents the current minimum value. Initially, a large number (999999) is assigned. $j+$ NLINES $+ 20$ is the row number on which the t-value is written. To delete the column with the smallest absolute value of t-values, write the following macro in the VBA program:

```
<Begin>
TRange = Chr(Asc("B") + IMIN - 1)
 Columns(TRange).Select
  Selection.Delete Shift:=xlToLeft
  CVARS = CVARS - 1
<End>
```

The explanatory variable data begin in Column B; therefore, select the IMINth explanatory variable data column (the data column name is the character-type data TRange), delete it, and execute Align Left. The column deletion also reduces the number of explanatory variables (CVARS) at the current stage by one.

The macro for executing Regression under the Data Analysis tools is as follows:

```
<Begin>
XRANGE = "B2:" + Chr(65 + NVARS - i + 1) + Trim(Str(NLINES + 2))

Application.Run "ATPVBAEN.XLAM!Regress",
    ActiveSheet.Range(YRANGE),_ActiveSheet.Range(XRANGE),
    False, True, , ActiveSheet.Range(OVERVIEW)_, False, False,
    False, False, , False
<End>
```

The "_" indicates that the program line continues; thus, the statement beginning with Application.Run spans three lines. User-specified information in the Regression wizard is listed. Thus, the checkboxes are determined by True/False to indicate whether they are checked or unchecked. YRANGE is the column of the objective variable, XRANGE is the column list of the explanatory variables, and OVERVIEW is the upper left cell of the data area where the results of the regression analysis will be listed.

Connecting the three parts shown as a program completes the variable reduction program. The completed version is shown below.

```
<Begin>
Sub reduce_variables()
NVARS = Cells(1, 2)
NLINES = Cells(1, 1)
criteria = Cells(1, 3)
If criteria < 1 Then criteria = 2
YRANGE = "A2:A" + Trim(Str(NLINES + 2))
OVERVIEW = "A" + Trim(Str(NLINES + 4))
CVARS = NVARS
For i = 1 To NVARS
XRANGE = "B2:" + Chr(65 + NVARS - i + 1) + Trim(Str(NLINES + 2))
    Application.Run "ATPVBAEN.XLAM!Regress",
        ActiveSheet.Range(YRANGE),_ActiveSheet.Range(XRANGE),
```

```
      False, True, , ActiveSheet.Range(OVERVIEW)_, False, False,
      False, False, , False
TMIN = 999999
  For j = 1 To CVARS
        TVAL = Abs(Cells(j + NLINES + 20, 4))
        If TVAL < TMIN Then
            TMIN = TVAL
            IMIN = j
        End If
  Next j
  If TMIN > criteria Then
     Exit For
  End If
TRange = Chr(Asc("B") + IMIN - 1)
Columns(TRange).Select
Selection.Delete Shift:=xlToLeft
CVARS = CVARS - 1
Next i
End Sub
<End>
```

The variable criteria are set for the absolute value of the t-value. If it is set to 2, then explanatory variables with t-values less than or equal to 2 in absolute value are deleted. The Trim function removes spaces before and after a character in character-type data.

2.4. What-if analysis

Excel's What-If Analysis tools allow you to set various conditions on your data and simulate results based on them. Excel provides two What-If Analysis tools. One is the "Data Table" and the other is the "Scenario Manager". This section will show you how to perform a what-if analysis using Data Table.

Prior to executing the what-if analysis, perform a regression analysis using temperature data. Express the average temperature data for 14 major cities in the Tohoku region in January 2000 in longitude and latitude. As we want to find out the minimum temperature, we use the squared data (x^2, y^2, xy) in addition to the longitude and latitude (xy) to express the temperature on a quadratic surface. That is, the temperature is calculated with five explanatory variables (x, y, x^2, y^2, xy) as follows:

$$T = a_0 + a_1 x + a_2 y + a_3 x^2 + a_4 y^2 + a_5 xy \tag{2.17}$$

Using the Data Analysis tools in Excel that we have already described, execute a regression analysis and calculate the coefficients $(a_0, a_1, a_2, a_3, a_4, a_5)$.

Enter the longitude (x) "140–142° (in 0.2° increments)" of the location for which you want to predict the temperature in the second and subsequent rows of Column J. Then, in the first row, Column J and onward, enter the latitude (y) "37–41.5° (in 0.2° increments)" of the location for which you want to predict the temperature. The data table uses a regression analysis model to visualize the temperature in a two-dimensional space based on latitude and longitude (Figure 2.21).

To use the Data Table, first select the cells (J1:U13) to be visualized. Next, go to the [Data] tab in Excel, click [What-If Analysis], and then select [Data Table]. The dialog box appears. Set the following:

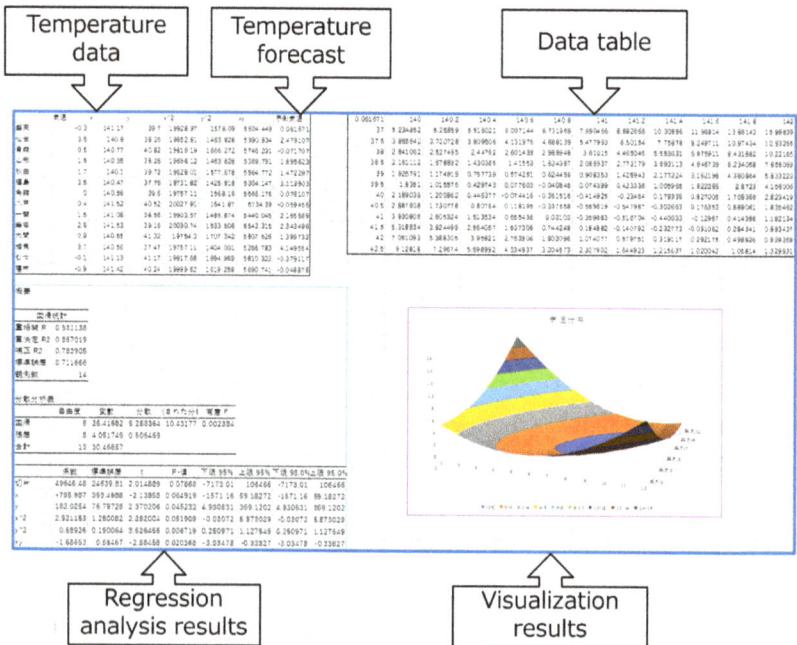

Figure 2.21. Visualization of regression analysis results using data table.

(1) Select cell C2 in [Row input cell], where the longitude (x) will be assigned.
(2) Select cell D2 in [Column input cell], where the longitude (y) will be assigned.
(3) Click [OK] to create a data table.

You can now see how the temperature data change with the combination of longitude and latitude.

To visualize the data table, Excel's conditional formatting and contour display are useful.

Conditional formatting allows you to automatically change the background or text color of a cell based on the value or data in a cell. This is called "conditional formatting". Conditional formatting can be used to display data in an easy-to-view manner. To use conditional formatting, follow the given steps:

(1) Select the cells you want to apply conditional formatting to.
(2) Select [Conditional Formatting] from the [Format] tab.
(3) The "Conditional Formatting" dialog box appears. Set the following:
(4) Click [Add] to enter conditions.
(5) In [Format], set the cell and text color to highlight when a condition is met.
(6) Click [OK] to complete the Conditional Formatting settings.

Conditional formatting allows you to set the formatting freely, for example, red if the cell value is greater than a specific value, blue if the cell value is less than a specific value, and gray otherwise. You can also set multiple conditions.

In Excel, you can create diagrams called "Surface" to visualize data. Surface is automatically color-coded according to the data, making the data easier to understand.

To create a Surface in Excel, select [Surface] from the [Data] tab. A "Chart Design" dialog box appears. Set the following:

(1) In [Data] settings, specify the range of data for which contour lines are to be created.
(2) In [Chart Styles], set the display method and color of the contour lines.
(3) Click [OK] to complete the creation of the surface.

The created surface is displayed on an Excel sheet. Contour lines can be used to visualize trends and patterns in the data. Contour lines can also be used to identify changes and fluctuations in the data.

2.5. Solver

The Excel Solver is an optimization tool included in Microsoft Excel. The Excel Solver can be used to solve optimization problems, including minimum, maximum, and constrained optimization problems. It can also be used to solve a wide variety of problems by selecting from various optimization algorithms.

The Excel Solver offers three methods (GRG Nonlinear, LP Simplex, and Evolutionary).

GRG nonlinear [20] is used to solve nonlinear optimization problems. GRG nonlinear solves nonlinear optimization problems using a method called Generalized Reduced Gradient (GRG). This method infers several optimal solutions to solve a problem and then selects the best one among them. GRG nonlinear is characterized by being extremely fast. On the other hand, it is said that the solution may not be exact. Therefore, if you seek high accuracy, you should consider using different algorithms.

A linear optimization problem refers to the problem of maximizing or minimizing an equation that linearly combines a variable with one or more coefficients.

For example, the following problem is a linear optimization problem:
Find x, y that maximizes the following objective function:

$$z = 3x + 4y \tag{2.18}$$

Find x, y that satisfy the following constraints:

$$x + y \leq 10$$

$$2x + y \geq 5$$

$$x, y \geq 0$$

To solve this problem, optimization methods are usually used to evaluate sample points and estimate the optimal solution using those points. Examples include linear programming and simplex methods.

Nonlinear optimization problems refer to problems in which variables are combined nonlinearly with coefficients to maximize or minimize an equation. For example, the following problem is a nonlinear optimization problem:

Find x, y that maximize the following objective function:

$$z = x^2 + y^2 \qquad (2.19)$$

Find x, y that satisfy the following constraints:

$$x^2 + y^2 \leq 10$$

$$2x + y \geq 5$$

$$x, y \geq 0$$

To solve this problem, the Newton or steepest descent method is usually used. Nonlinear optimization problems can be difficult to solve because there may be no solution or multiple solutions.

LP Simplex [21] is used to solve linear optimization problems. LP Simplex solves linear optimization problems using a method called the Linear Programming (LP) Simplex Method, which optimizes the values of variables to satisfy an objective function and constraints. LP Simplex is characterized by being extremely fast. It is also characterized by the accuracy of its solution. On the other hand, if the problem is complex, you should consider using other algorithms.

A linear optimization problem refers to the problem of maximizing or minimizing an equation that linearly combines a variable with one or more coefficients.

For example, the following problem is a linear optimization problem:
Find x, y that maximize the following objective function:

$$z = 3x + 4y \qquad (2.20)$$

Find x, y that satisfy the following constraints:

$$x + y \leq 10$$

$$2x + y \geq 5$$

$$x, y \geq 0$$

To solve this problem, optimization methods are usually used to evaluate sample points and estimate the optimal solution using those points. Examples include linear programming and simplex methods.

Evolutionary [22] is used to solve nonlinear optimization problems. Evolutionary uses a method called Evolutionary Algorithm (EA) to solve nonlinear optimization problems. This method mimics the genetic process in nature, in which multiple optimal solutions are estimated, and the best one is selected to solve the problem. Evolutionary is characterized by the fact that it takes longer to compute than other algorithms. On the other hand, when the problem is complex, it can produce better results than other algorithms.

Here is an example of using the Solver to explore an extremely cold region. We will use the temperature data used in the regression analysis to perform the optimization calculations. First, prepare for the optimization on row 16. Enter "Extreme Cold Region" as the location's name in A16 (not mandatory). Enter the initial values for longitude and latitude in C16 and D16 as 141.0 and 40.0, respectively. Copy H15, the cell showing the predicted temperature, and paste it into H16. Confirm that the predicted temperatures are calculated by the regression model shown in H16.

Next, start the Excel Solver. Click the [Data] tab in Excel and click [Solver]. In the Solver Parameters dialog box, select "H16" in [Set Objective]. In "To", check [Min] as an objective value. And in [By Changing Variable Cells], select the two cells C16:D16 where the longitude and latitude are input as variable cells, and then click the [Solve] button. After a period of time, a dialog box will appear indicating that the optimization is complete. Press the [OK] button to save the optimization results. Confirm that the minimum value is shown in H16. To find out where the extremely cold region is, you can use the latitude and longitude to search for it on Google Maps (Figure 2.22).

2.5.1. *Optimization*

Optimization means finding the best solution to solve a problem with a certain objective. Methods for solving such problems are called "optimization algorithms". Optimization problems are commonly used in a variety of fields, including computer science, engineering, and economics.

For example, a well-known optimization problem is finding the minimum value. In this problem, the objective is to find an input that minimizes the value of a function. The same is true for the problem of finding the maximum value. In this problem, the objective is to find the input that maximizes the value of a function.

In some optimization problems, conditions may also need to be satisfied. An example is when a program needs to be minimized but at the same time must meet the minimum specifications required to process a particular set of data. Such a problem is called a "constrained optimization problem".

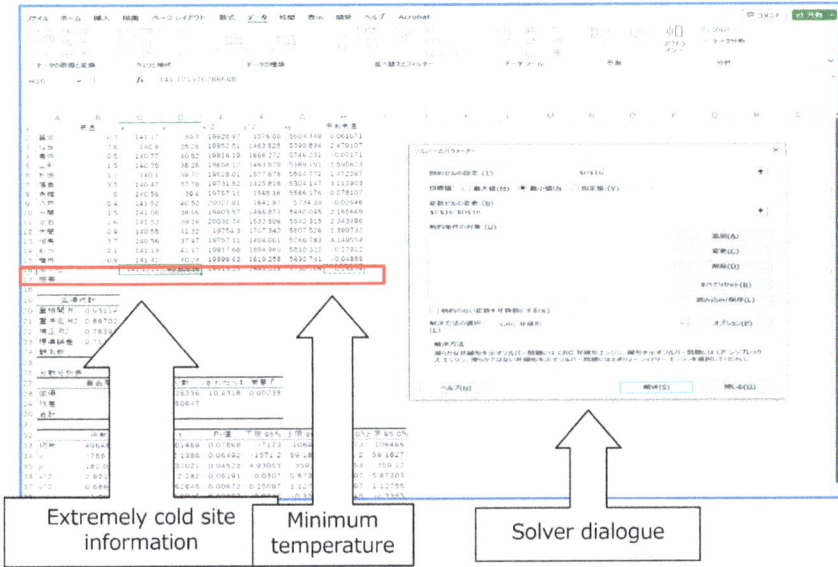

Figure 2.22. Solver calculations for extremely cold region.

2.5.2. *Implementation of regression analysis*

We have discussed the regression analysis using the Data Analysis tools in Excel. The regression analysis can also be performed using a Solver.

Now, we determine an error in a simple regression equation that represents a regression line. Assume that the equation of the regression line that can successfully account for N discrete points is $y = \beta_0 + \beta_1 x$. For the ith point, the error in the y axis direction is $e_i = y_i - (\beta_0 + \beta_1 x_i)$. The sum of squares of this error is

$$\sum_{i=1}^{N} e_i^2 = \sum_{i=1}^{N}(y_i - (\beta_0 + \beta_1 x_i))^2 \tag{2.21}$$

(Figure 2.23).

To calculate the coefficients (β_0, β_1) of the equation that minimizes the error, we use the Solver as follows (Figure 2.24):

First, enter the five discrete data points into cells A2:B6 on the Excel sheet. Then, enter the formula for the square of the error as '=($D$2*A2+ $E$2-B2)^2' into cell C2. The slope and intercept $((\beta_0, \beta_1))$ of the regression line are set to the initial values (D2, E2) = (1,1) and are absolutely

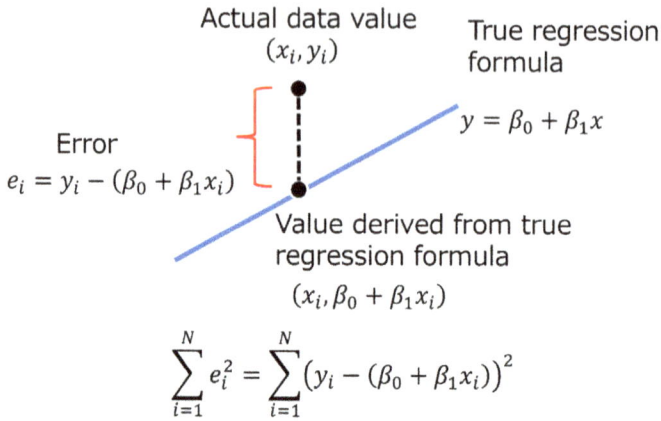

Figure 2.23. The sum of the squares of the error in the regression line.

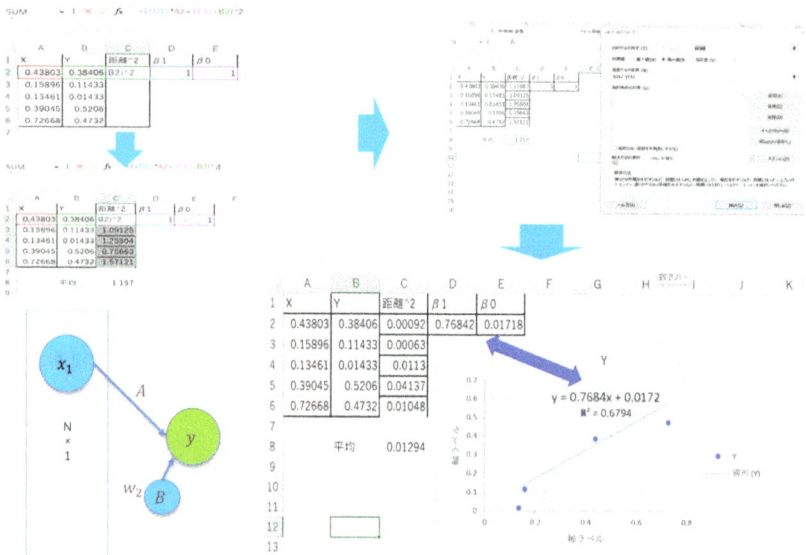

Figure 2.24. Identification of regression lines by solvers.

referenced. In contrast, the coordinate values (A2, B2) are referenced relative to each other. Next, copy four rows of the square of the error in C2. Write the formula for finding the mean of the squares of the errors for the five rows in C8. Execute Solver by following the given steps:

(1) Click the [Data] tab in Excel and click [Solver].
(2) In the Solver Parameters dialog box, select "C8" in [Set Objective].
(3) For "To" in the same dialog box, select [Min] as an objective value.
(4) In the same dialog box, set "D2:E2" in [By Changing Variable Cells].
(5) In the same dialog box, for [Subject to the Constraints], no constraints are required to identify the regression line.
(6) In the same dialog box, uncheck [Make Unconstrained Variables Non-Negative].
(7) In the same dialog box, select [GRG Nonlinear] in Select a Solving Method.
(8) In the same dialog box, press the [Solve] button.

In the Solver Results dialog box, check [Keep Solve Solution], and then click the [OK] button.

The slope and intercept calculated by the Solver are now obtained from the coordinate data of the 5 points. When comparing these results to the slope and intercept of the regression line displayed in the Excel scatter plot, you will note that the results are roughly consistent.

For a regression plane that can successfully account for N discrete points, we first consider the error in the multiple regression equation. The equation of the regression plane that can successfully account for N discrete points is found by minimizing the mean square error

$$\sum_{i=1}^{N} e_i^2 / N = \sum_{i=1}^{N} (z_i - (\beta_0 + \beta_1 x_i + \beta_2 y_i))^2 / N. \qquad (2.22)$$

To show a concrete example, let us find the regression plane from the coordinate data of $N = 26$ points.

First, enter the 26 discrete point data into the cell range "B2:C27" in the Excel sheet. Column B represents X, and Column C represents Y. These coordinate data are randomly selected from $\{(x, y) | 0 \le x \le 1, 0 \le y \le 1\}$. For X and Y, calculate the value of Z using the following formula and enter it in Column A:

$$z = x^2 + x + y^3 + y^2 \qquad (2.23)$$

Next, click [Data Analysis] — [Regression]. Select Columns B and C in [Input X Range] as the explanatory variables and Column A in [Input Y Range] as the objective function. Select G2 in [Output Range] (refer to the top part of Figure 2.25). Check whether the coefficients obtained at this time (intercept, coefficient of X, and coefficient of Y) are the same as those obtained by minimizing the mean square error with the Solver.

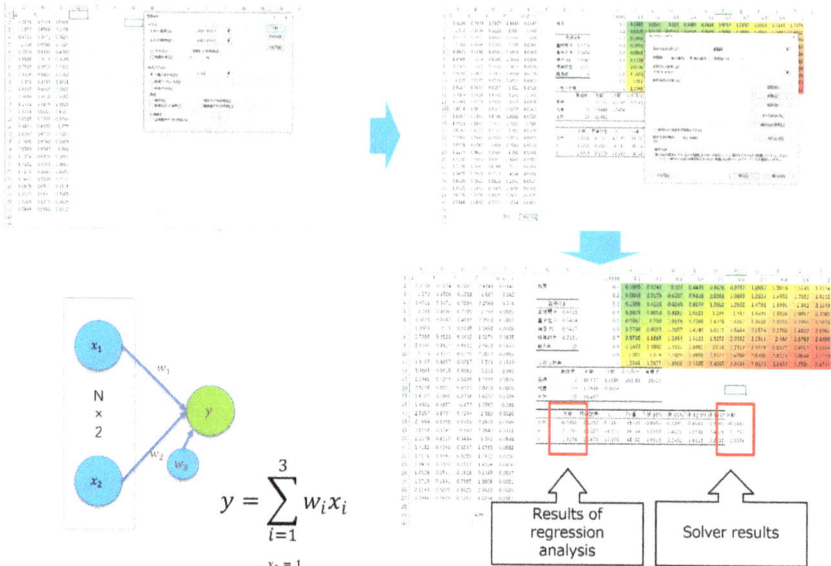

Figure 2.25. Identification of regression planes by solver.

Calculate the following using the coefficients tentatively determined for D2 (stored in P18:P20) and the coordinate data (stored in B2:C2) and label the obtained value as Z':

$P$18+MMULT (B2:C2, $P$19:$P$20)

Write the formula '(A2-D2)^2' for the square of the error in E2. Next, copy 25 rows of E2 for the square of the error. Then, write the formula for the mean of the square of the error for 26 rows into E29.

Execute Solver by following the given steps:

(1) Click the [Data] tab in Excel and click [Solver].
(2) In the Solver Parameters dialog box, select "E29" in [Set Objective].
(3) For "To" in the same dialog box, select [Min] as an objective value.
(4) In the same dialog box, set "P18:P20" in [By Changing Variable Cells].
(5) In the same dialog box, for [Subject to the Constraints], no constraints are required to identify the regression line.
(6) In the same dialog box, uncheck [Make Unconstrained Variables Non-Negative].

(7) In the same dialog box, select [GRG Nonlinear] in Select a Solving Method.
(8) In the same dialog box, press the [Solve] button.

In the Solver Results dialog box, check [Keep Solve Solution], and then click the [OK] button.

The coefficients calculated by the Solver are now obtained from the coordinate data of 26 points. That is, the Solver was used to create the model with linear regression from $N = 26$ discrete point data (X, Y), selected at random from $\{(x, y)|0 \le x \le 1, 0 \le y \le 1\}$. The mean squared error was 0.0402. When comparing these results with the coefficients obtained from the regression analysis in the Excel analysis tool, you will note that they are roughly consistent (refer to the bottom part of Figure 2.25).

2.6. Colab

Colab allows you to read data and perform statistical analysis by using Pandas and NumPy, Python's libraries for data analysis. For example, you can use Pandas to read data from a CSV file and calculate the mean, standard deviation, minimum, and maximum values as follows:

```
import pandas as pd

df = pd.read_csv("sample_data.csv")
mean = df["column"].mean ()
std = df["column"].std ()
min = df["column"].min ()
max = df["column"].max ()
```
NumPy can similarly read data and perform statistical analysis.

Colab also allows you to visualize the data using Matplotlib or Seaborn. For example, you can use Matplotlib to display a histogram of your data as follows:

```
import matplotlib.pyplot as plt

plt.hist(df["column"])
plt.show()
```

2.6.1. *Correlation analysis*

To perform correlation analysis in Colab, it is convenient to use Pandas. In Pandas, the "corr" method can be used to easily calculate correlation coefficients for variables in a data frame. For example, you can use Pandas to compute the correlation coefficient between two variables as follows:

import pandas as pd

```
df = pd.read_csv("sample_data.csv")
corr = df["column1"].corr(df["column2"])
print(corr)
```

The code calculates the correlation coefficient between "column1" and "column2" and assigns the result to the variable "corr". In Pandas, the "corr" method can also be used to specify the type of correlation coefficient. For example, to calculate the Pearson correlation coefficient, specify "method="pearson"".

```
corr = df["column1"].corr(df["column2"], method="pearson")
```

Pandas also allows you to calculate the correlation coefficient for all variables in a data frame by using the "corr" method.

```
corr_matrix = df.corr()
print(corr_matrix)
```

The code calculates the correlation coefficients for all variables in the data frame "df" and assigns the results to the variable "corr_matrix".

2.6.2. *F-test*

To perform F-tests in Colab, it is convenient to use the SciPy library. The SciPy library provides an "f_oneway" function that can be used to perform F-tests. The F-test examines whether the means of a variable in groups of three or more are significantly different. For example, the F-test can be performed using SciPy as follows:

from scipy import stats

```
group1 = [1, 2, 3, 4]
group2 = [2, 3, 4, 5]
```

```
group3 = [3, 4, 5, 6]

stat, p = stats.f_oneway(group1, group2, group3)
print ("F statistic:", stat)
print ("p-value:", p)
```

The code prepares three groups and uses the "f_oneway" function to perform an F-test. F statistic and p-value results are obtained from the F-test. If the F statistic is large, we can conclude that there is a significant difference in means between the groups. On the other hand, if the p-value is small, we can conclude that there is a significant difference between the means of the groups.

2.6.3. *t-Test*

To perform t-tests in Colab, it is convenient to use the SciPy library. The SciPy library provides a "ttest_ind" function that can be used to perform t-tests. The t-test examines whether two groups differ significantly in the mean value of a variable. For example, the t-test can be performed using SciPy as follows:

```
from scipy import stats

group1 = [1, 2, 3, 4]
group2 = [2, 3, 4, 5]

stat, p = stats.ttest_ind(group1, group2)
print ("t statistic:", stat)
print ("p-value:", p)
```

The code prepares two groups and uses the "ttest_ind" function to perform a t-test. t statistic and p-value are obtained as the result of the t-test. On the other hand, if the p-value is small, we can conclude that there is a significant difference in the means between the groups. On the other hand, if the p-value is small, we can conclude that there is a significant difference between the means of the groups.

In Colab, if the variances are considered to be equal between the two groups, it is convenient to use the SciPy library to perform the t-test. The SciPy library provides a "ttest_ind" function that can be used to perform t-tests. For example, we can use SciPy to perform a t-test to see if the variances can be considered equal between the two groups as follows:

```
from scipy import stats

group1 = [1, 2, 3, 4]
group2 = [2, 3, 4, 5]

stat, p = stats.ttest_ind(group1, group2, equal_var=True)
print ("t statistic:", stat)
print ("p-value:", p)
```

This code uses a "ttest_ind" function to perform a t-test when the variance cannot be considered equal between the two groups. If the "equal_var" parameter is set to "False", a t-test can be performed if the variances are not considered equal between the two groups.

In Colab, it is convenient to use the SciPy library to examine whether the means of some variables in the same group are significantly different. The SciPy library provides a "ttest_rel" function that can be used to test for correlation. For example, you can use SciPy to test whether the means of some variables in the same group are significantly different.

```
from scipy import stats

group = [1, 2, 3, 4]

stat, p = stats.ttest_rel(group, group)
print ("t statistic:", stat)
print ("p-value:", p)
```

The code uses a "ttest_rel" function to test whether the means of some variables in the same group are significantly different. t statistic and p-value are obtained as the result of the t-test. On the other hand, if the p-value is small, we can conclude that there is a significant difference in the means between the groups. On the other hand, if the p-value is small, we can conclude that there is a significant difference between the means of the groups.

2.6.4. *Z-test*

To perform *Z*-tests in Colab, it is convenient to use the SciPy library. The SciPy library provides a "zscore" function that can be used to perform *Z*-tests. The *Z*-test examines whether the sample means are significantly different, given that the mean of a population is known. For example, the *Z*-test can be performed using SciPy as follows:

```
from scipy import stats

sample = [1, 2, 3, 4]
population_mean = 2.5

z_score = stats.zscore(sample, population_mean)
print(z_score)
```

This code uses the "zscore" function to compute the Z-value of the sample. If the Z-value is large, we can conclude that the sample means are significantly different.

2.6.5. *Regression analysis*

To perform regression analysis in Colab, it is convenient to use the NumPy or SciPy library. For example, a single regression analysis can be performed using NumPy as follows:

```
import numpy as np

# Explanatory variables
x = np.array([1, 2, 3, 4])
# Objective variables
y = np.array([1, 2, 3, 4])

# Find regression coefficients
a, b = np.polyfit(x, y, 1)
print ("regression coefficient:", a)
print ("intercept:", b)
```

This code uses the "polyfit" function to find the regression coefficient a and intercept b of the regression line between the explanatory variable x and the objective variable y. The SciPy library can also be used to perform multiple regression analysis, logistic regression analysis, etc.

To perform regression analysis in Colab, it is convenient to use the NumPy or SciPy library. For example, a single regression analysis can be performed using NumPy as follows:

```
import numpy as np

# Explanatory variables
x = np.array([1, 2, 3, 4])
# Objective variables
y = np.array([1, 2, 3, 4])

# Find regression coefficients
a, b = np.polyfit(x, y, 1)
print ("regression coefficient:", a)
print ("intercept:", b)
```

This code uses the "polyfit" function to find the regression coefficient a and intercept b of the regression line between the explanatory variable x and the objective variable y. The SciPy library can also be used to perform multiple regression analysis, logistic regression analysis, etc.

In Colab, it is convenient to use the SciPy library to calculate t-values and p-values for regression coefficients in regression analysis. The SciPy library provides a "linregress" function that can be used to obtain the results of a regression analysis. For example, you can use SciPy to obtain the results of a regression analysis as follows:

```
import numpy as np
from scipy import stats

# Explanatory variables
x = np.array([1, 2, 3, 4])
# Objective variables
y = np.array([1, 2, 3, 4])

# Find regression coefficients
slope, intercept, r_value, p_value, std_err = stats.linregress(x, y)
```

```
print ("regression coefficient:", slope)
print ("intercept:", intercept)
print ("t value:", slope / std_err)
print ("p value:", p_value)
```

This code uses the "linregress" function to find the regression coefficient, intercept, t-value, p-value, and standard error of the regression line between the explanatory variable x and the objective variable y.

In Colab, it is convenient to use the StatsModels library to calculate t-values and p-values for regression coefficients in multiple regression analysis. The StatsModels library provides an "OLS" class that can be used to perform multiple regression analysis. For example, you can use StatsModels to perform multiple regression analysis as follows:

```
import statsmodels.api as sm

# Explanatory variables
x = np.array([[1, 2], [3, 4], [5, 6], [7, 8]])
# Objective variables
y = np.array([1, 2, 3, 4])

# Perform multiple regression analysis
model = sm.OLS(y, x)
results = model.fit()

# Find regression coefficients
print(results.summary())
```

This code uses the "OLS" class to perform multiple regression analysis. The "fit" method can be used to obtain the results of the regression analysis. The "summary" method allows you to see the results of the multiple regression analysis, including t-values and p-values for the regression coefficients.

The following is a sample code that reads an Excel file in Colab and performs a regression analysis:

```
<Begin>
# Import Pandas
import pandas as pd
# Import files from Google Drive
from google.colab import drive
```

```
drive.mount('/content/drive')
Excel file path
file_path = '/content/drive/My Drive/data.xlsx'
Read Excel files
df = pd.read_excel(file_path)
# Import necessary libraries
import numpy as np
from sklearn.linear_model import LinearRegression
# Set explanatory and objective variables
X = df[['x1', 'x2', 'x3']]
y = df['y']
# Generate a linear regression instance
model = LinearRegression()
# Learn
model.fit(X, y)
# Coefficients
print ("coefficient:", model.coef_)
# Intercept
print ("intercept:", model.intercept_)
# Coefficient of determination
print ("R2:", model.score(X, y))
<End>
```

This sample uses Pandas to read an Excel file and uses sklearn's Linear-Regression to perform a regression analysis. The above code reads the file from Google Drive. In order to access Google Drive, we need to authenticate it using drive.mount(). Also, in the above-mentioned example, $x1$, $x2$, and $x3$ are set as explanatory variables, and y is the objective variable.

2.6.6. *Optimization problem*

To solve optimization problems with Colab, it is convenient to use the SciPy library. The SciPy library provides several functions for solving optimization problems. By using them, you can solve optimization problems. For example, you can use SciPy to solve the optimization problem of finding the minimum value as follows:

```
from scipy import optimize

# Function f(x) = x^2
```

```
def f(x):
    return x**2

# Find the minimum value
result = optimize.minimize(f, x0=0)
print(result)
```

This code uses the "minimize" function to find the minimum value of the function $f(x) = x^2$. There are also several types of optimization problems. For example, there are problems with finding the minimum and maximum values and problems with constraints. The SciPy library provides functions for each of these types. By using the corresponding functions, you can solve optimization problems.

2.7. NNs

This document uses a regression model to build a spatiotemporal model from physical data. In this model, regression tries to match the given physical data as accurately as possible, but unlike interpolation, the regression model is not a perfect fit for the data. It does, however, learn features from the data. In this section, we employ NNs as the regression method. While various structures have been proposed for NNs, we focus here on FCNN (Fully Connected NN) (Figure 2.26). An FCNN consists of an input layer, a hidden layer, and an output layer. Each layer has a specific number of neurons. The output value of each neuron is calculated by the inner product of a vector consisting of the input values and a vector consisting of the weights. Therefore, the mapping from the input layer to the output layer ultimately yields the desired output value. An FCNN is a neural network in which every neuron is connected to all neurons in the previous layer. It is also called a densely connected neural network, a fully connected neural network, or a multilayer perceptron (MLP).

FCNNs have a large number of weight parameters between the input and output because each layer is connected to all neurons in the immediately preceding layer. This structure allows it to hold a very large number of parameters from the input to the output layer and, therefore, has a high expressive capability. However, a large number of layers or neurons can cause slow learning or overlearning problems. However, if the number of layers or the number of neurons is large, learning can be slow or cause overlearning problems. In this document, NN shall mean FCNN.

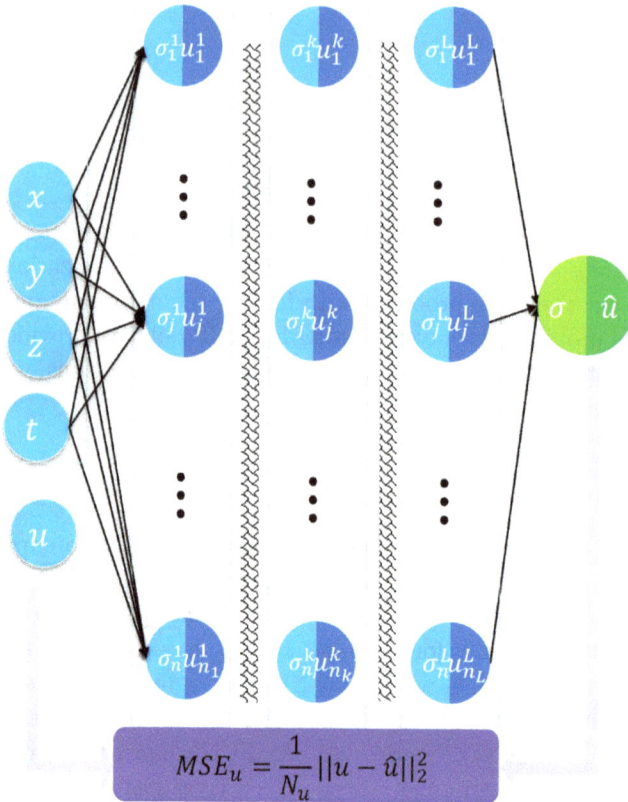

Figure 2.26. *L*-layer neural networks.

In neural networks, learning is the process of automatically adjusting parameters such as weights and biases of a neural network using training data. Through this process, the neural network builds a model that is optimized by the training data.

Neural network training is divided into two main steps: "forward propagation" and "back propagation". Forward propagation is a step in which the process from the input layer to the output layer is executed in sequence, with the input data passing through the network to produce the output. Back propagation is the step in which the error is back propagated from the output layer to the input layer, updating weights and biases to minimize the error.

Specifically, for training, a measure called the loss function is defined, and weights and biases are updated to minimize the difference between the neural network's output and the ground truth data. Therefore, the neural network is adjusted to fit the training data. Regularization techniques, dropouts, and other methods may also be used to prevent overlearning.

Once sufficiently trained, neural networks can achieve high predictive accuracy for unknown data. Learning in neural networks is a type of optimization. Learning in neural networks aims to obtain optimal prediction results by using training data and adjusting parameters such as weights and biases. Now, assume that the NN consists of L layers and that n_i neurons are defined in layer i. The number of parameters to be optimized can be calculated as follows:

$$\sum_{i=1}^{L+1} (n_i + 1)n_{i+1} \tag{2.24}$$

In this equation, $i = 1$ represents the input variable, so $n_1 = 4$ for spacetime coordinates. The addition of 1 is due to the inclusion of bias as a type of weight. Also, $i = L + 2$ represents the output variable, so $n_{L+2} = 1$ for scalar data. Figure 2.26 shows an NN consisting of L layers.

Weights are defined on arcs connected to the neurons and output variables. For small NNs, we focus on clarity and explain how to optimize the weights using the Excel Solver. For practical use, we will use keras and other libraries introduced in Chapter 1.

The learning process searches for parameter values that minimize the loss function in order to minimize the difference between the output of the neural network and the ground truth data. This search process is performed using optimization algorithms (e.g., gradient descent method).

As the learning progresses, the parameters are updated, and the value of the loss function is reduced to obtain the optimal parameters. Using these optimal parameters, the neural network can achieve high prediction accuracy for unknown data.

2.7.1. *Universal approximation theorem*

The universal approximation theorem is a theorem that states that an NN can approximate any continuous function with any precision. In other words, a properly designed NN can approximate a desired output for any given input.

This theorem is one of the reasons why NNs have become widely used and is the basis for demonstrating their usefulness. In practice, however, it is not always easy to achieve, as it requires a sufficient number of neurons, appropriate algorithms, datasets, etc., to be able to approximate with any precision that the theorem suggests.

The universal approximation theorem was first proposed in 1989 in the paper "Multi-Layer Feedforward Networks are Universal Approximators" by George Cybenko and Edward G. Widrow [23]. This paper was published in the IEEE Transactions on Neural Networks and Learning Systems and later became an important foundation for the theory of neural networks. In this paper, George Cybenko and Edward G. Widrow proved that there exist neural networks that can approximate any continuous function with any precision. They showed that this theorem made neural networks very powerful and led to the widespread use of neural networks.

The existence of a neural network that can approximate any continuous function with any precision can be proved as follows:

- First, a neural network with one single sigmoid function can approximate a Lipschitz continuous function. A Lipschitz continuous function is a function with a rate of change bounded by some constant, and it is possible to approximate such a function.
- Second, a neural network with multiple sigmoidal functions can approximate any Lipschitz continuous function. This is because each layer takes the previous layer's output as input, thus allowing more complex functions to be represented.
- Finally, any continuous function can be represented as the sum of an infinite number of Lipschitz continuous functions. This means that a neural network with multiple sigmoidal functions can be used to approximate any continuous function with any precision.

Therefore, a Lipschitz continuous function is a continuous function in which a constant L exists, and the following equation holds for the function f:

$$| f(x) - f(y) | \leq L \, | \, x - y \, | \tag{2.25}$$

In this equation, x and y represent any two points within the domain of the definition of f, and $| \cdot |$ represents the absolute value. This definition indicates that "the change in the value of f is bounded in proportion to the change in its argument". In other words, the closer x and y are to each other, the smaller the difference between $f(x)$ and $f(y)$.

Lipschitz continuous functions play an important role in calculus, functional analysis, and other areas of mathematics. They are also used in various fields, such as physics, engineering, and economics.

As shown above, a neural network with one sigmoidal function approximates a Lipschitz continuous function, and a neural network with multiple sigmoidal functions can approximate any Lipschitz continuous function. Therefore, any continuous function can be approximated with any precision.

2.7.2. *Regression analysis using NNs*

To explain the regression analysis using NNs, we use as discrete data the bivariate data described in Section 5, paragraph 2. These data are calculated from the bivariate function

$$y = x_1^2 + x_1 + x_2^3 + x_2^2. \tag{2.26}$$

The mean squared error of the model

$$y = w_0 + w_1 x_1 + w_2 x_2, \tag{2.27}$$

created by linear regression from 26 discrete data randomly selected from $\{(x_1, x_2)|0 \le x_1 \le 1, 0 \le x_2 \le 1\}$, was 0.0402. Using the coefficients $w_0 w_1 w_2$ as parameters, the parameters that minimize the mean square error between the spatiotemporal model and the given discrete data were calculated using the Excel Solver function.

In the previous section, we discussed specific ways to achieve regression analysis using the Solver. In this section, we will use the same data to explain how to achieve an NN model by using Solver. A simple NN model can be realized as an activation function connected to the output of this linear regression model. This time, we use the Sigmoid function

$$\sigma(x) = \frac{1}{1 + e^{-x}} \tag{2.28}$$

and assume that the output obtained using this function in the Excel sheet function is scale converted to

$$y = (y_{\max} - y_{\min}) \frac{1}{1 + e^{-(w_0 + w_1 x_1 + w_2 x_2)}} + y_{\min} \tag{2.29}$$

$\Delta y = (y_{\max} - y_{\min})$ to calculate the mean squared error between $y = \frac{\Delta y}{1 + e^{-(w_0 + w_1 x_1 + w_2 x_2)}} + y_{\min}$ and the given discrete data y_i.

$$\sum_{i=1}^{26} \left(y_i - \frac{\Delta y}{1 + e^{-(w_0 + w_1 x_{1i} + w_2 x_{2i})}} - y_{\min} \right)^2 \tag{2.30}$$

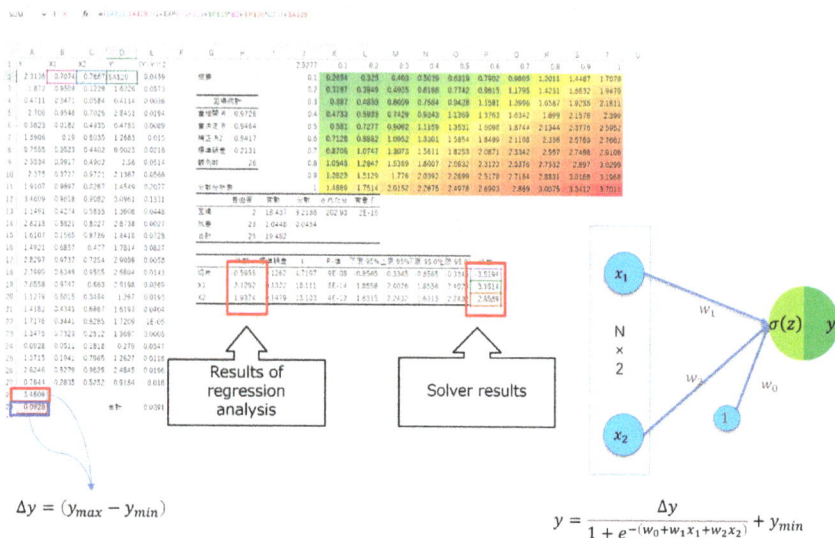

Figure 2.27. NN (Activation function only) model calculation using solver.

When the parameters (w_0, w_1, w_2) that minimize the mean squared error are calculated using the Excel Solver function, we note that the mean squared error is 0.0391, a slight decrease. However, this is not the result of the regression analysis (Figure 2.27).

This is a simple NN model with no intermediate layer. When the output,

$$y = \frac{\Delta y}{1 + e^{-(w_{11}^2 + w_{21}^2 m_1 + w_{31}^2 m_2)}} + y_{\min} \qquad (2.31)$$

$$m_1 = \frac{1}{1 + e^{-(w_{11}^1 + w_{21}^1 x_1 + w_{31}^1 x_2)}} \quad m_2 = \frac{1}{1 + e^{-(w_{12}^1 + w_{22}^1 x_1 + w_{32}^1 x_2)}},$$

obtained by adding one layer consisting of two variables (m_1, m_2), and the parameter $(w_{11}^1, w_{21}^1, w_{31}^1, w_{12}^1, w_{22}^1, w_{32}^1, w_{11}^2, w_{21}^2, w_{31}^2)$, which minimizes the mean squared error with the given discrete data, is calculated using the Excel Solver, the mean squared error is 0.01244, which is an even more significant reduction. You can note that the number of parameters increases from 3 to 9 when one intermediate layer is added (Figure 2.28). The activation function $\sigma(x)$ is listed on the left for the notation of neurons arranged in a semicircle. The variable m_1, m_2, y that receives the result is listed on the right, and its background color is changed.

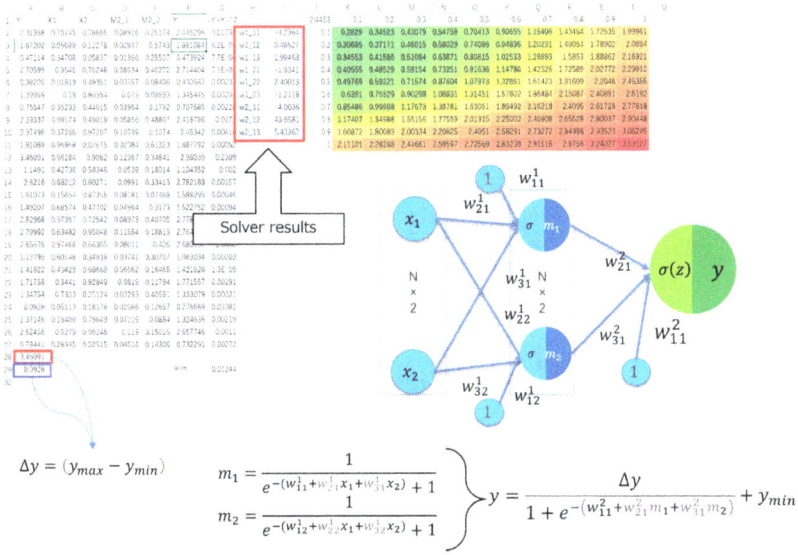

Figure 2.28. NN (Middle layer 1) model calculation using solver.

We can infer that the superiority of the NN model is due to the use of this activation function. The network structure corresponding to Equation (2.29) is shown in Figure 2.27. For Equation (2.31), a three-dimensional representation in the domain of definition $\{(x_1, x_2)|0 \leq x_1 \leq 1, 0 \leq x_2 \leq 1\}$ is shown in Figure 2.28.

2.7.3. NN implementation using Excel VBA

NN implementation using Excel allows you to build NNs with any structure. The structure (e.g., 4 neurons in layer 1, 6 neurons in layer 2, and 2 neurons in layer 3) can be entered manually in an Excel sheet. However, manual input has its limitations when entering more general structures. Therefore, we will discuss how to achieve a general multilayer perceptron using Excel VBA.

The aforementioned NN is executed using Excel VBA. Hereafter, we will refer to this program as ExNN. In this implementation, the NN is built by giving any number of layers and any number of neurons per layer. The weights that minimize the mean squared error calculated between the given objective variable and the prediction result by the NN are optimized by the Excel Solver.

First, assume that on Excel Sheet 1, the objective variable data is popu-lated in the first column, starting from the second row and continuing for as many observations as there are. Also, assume that the explanatory variable data is populated in the second and subsequent columns, starting from the second row, with a number of explanatory variables equal to the number of observations. Assume that labels for variables, etc., are entered in the first row, and sheet functions that calculate data values for explanatory variable data and neuronal variables are entered in the second and subsequent rows. Figures 2.29 and 2.30 show examples when the number of explanatory vari-ables is 2 and the number of observations is 26. Figure 2.31 shows an NN (hidden layer 2) model.

The first row contains the labels Y, $X\_i(i = 1$, number of explanatory variables), and $Mi\_j(i = 2$, number of layers, $j = 1$, number of neurons per layer) to identify the objective variable, explanatory variables, and neurons, respectively. Assuming that layer 1 consists of explanatory vari-ables, the label of the first neuron to enter along the row direction is M2_1 (Figure 2.30).

The second-row cell contains a sheet function that calculates the value of a neuron in each column for each layer according to a meta-parameter

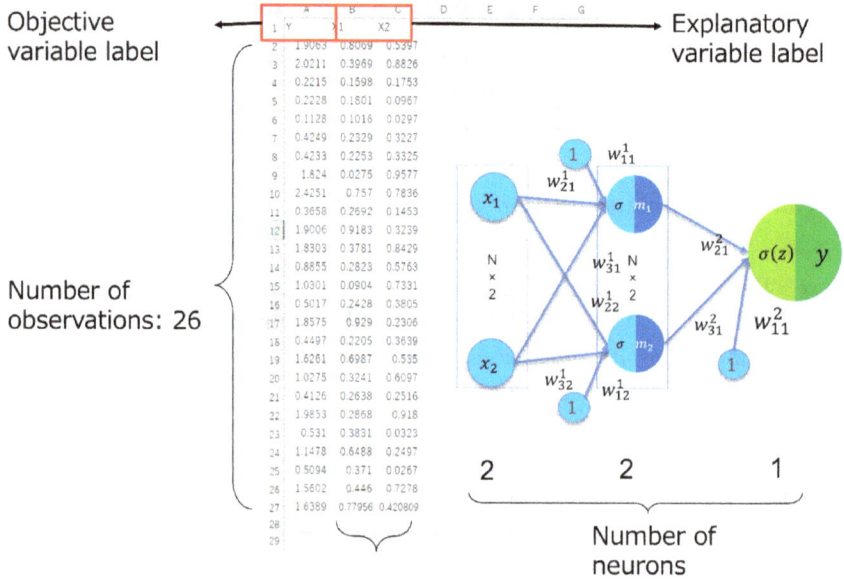

Figure 2.29. Target analysis data.

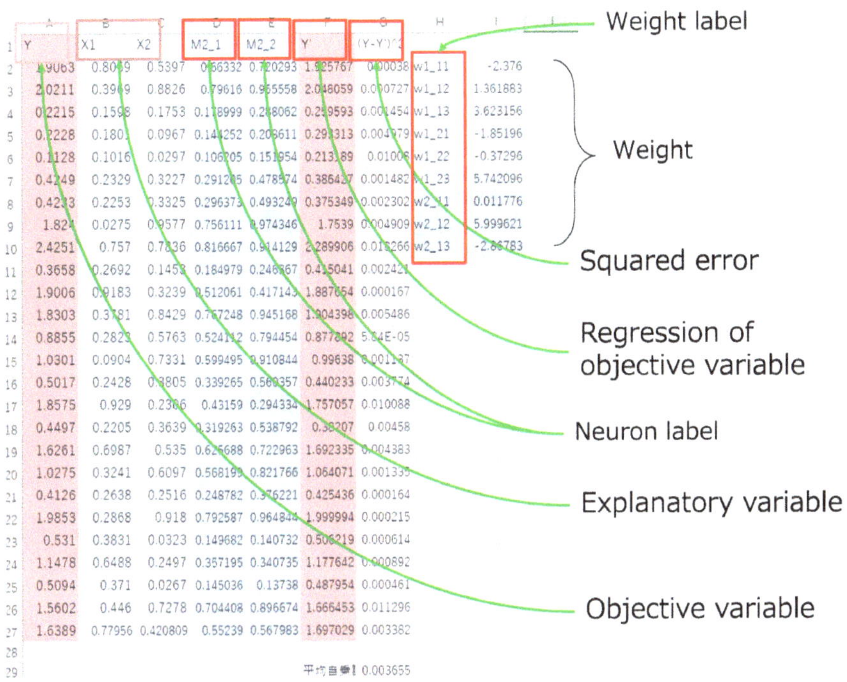

	Y	X1	X2	M2_1	M2_2	Y'	(Y-Y')²		
2	1.9063	0.8009	0.5397	0.66332	0.720293	1.925767	0.00038	w1_11	-2.376
3	2.0211	0.3909	0.8826	0.79616	0.965558	2.048059	0.000727	w1_12	1.361883
4	0.2215	0.1598	0.1753	0.118999	0.268062	0.259593	0.003454	w1_13	3.623156
5	0.2228	0.1801	0.0967	0.144252	0.205611	0.291313	0.004979	w1_21	-1.85196
6	0.1128	0.1016	0.0297	0.106705	0.151954	0.213189	0.01006	w1_22	-0.37296
7	0.4249	0.2329	0.3227	0.291205	0.478574	0.386427	0.001482	w1_23	5.742096
8	0.4233	0.2253	0.3325	0.296373	0.493249	0.375349	0.002302	w2_11	0.011776
9	1.824	0.0275	0.9577	0.756111	0.974346	1.7539	0.004909	w2_12	5.999621
10	2.4251	0.757	0.7836	0.816667	0.914129	2.289906	0.018266	w2_13	-2.86783
11	0.3658	0.2692	0.1453	0.184979	0.246667	0.415041	0.002421		
12	1.9006	0.9183	0.3239	0.512061	0.41714	1.887654	0.000167		
13	1.8303	0.3781	0.8429	0.757248	0.945168	1.904398	0.005486		
14	0.8855	0.2823	0.5763	0.524112	0.794454	0.877892	5.54E-05		
15	1.0301	0.0904	0.7331	0.599495	0.910844	0.99638	0.001137		
16	0.5017	0.2428	0.3805	0.339265	0.509357	0.440233	0.003774		
17	1.8575	0.929	0.2306	0.43159	0.294334	1.757057	0.010088		
18	0.4497	0.2205	0.3639	0.319263	0.538792	0.38207	0.00458		
19	1.6261	0.6987	0.535	0.625688	0.722963	1.692335	0.004383		
20	1.0275	0.3241	0.6097	0.568199	0.821766	1.064071	0.001335		
21	0.4126	0.2638	0.2516	0.248782	0.376221	0.425436	0.000164		
22	1.9853	0.2868	0.918	0.792587	0.964844	1.999994	0.000215		
23	0.531	0.3831	0.0323	0.149682	0.140732	0.506219	0.000614		
24	1.1478	0.6488	0.2497	0.357195	0.340735	1.177642	0.000892		
25	0.5094	0.371	0.0267	0.145036	0.13738	0.487954	0.000461		
26	1.5602	0.446	0.7278	0.704408	0.896674	1.666453	0.011296		
27	1.6389	0.77956	0.420809	0.55239	0.567983	1.697029	0.003382		
29				平成自乗1	0.003655				

Labels (right side): Weight label · Weight · Squared error · Regression of objective variable · Neuron label · Explanatory variable · Objective variable

Figure 2.30. Labels to be assigned to data.

for the number of neurons. In this sheet function, the labels (wi_jk, $i = 1$, number of neurons in layer $i + 1, j = 1$, number of neurons in layer $i + 1, k = 1$, number of neurons in layer i) and the values of the weights, which are defined for each neuron-to-neuron bond between layers, are filled in the column direction. The neuron values in layer $i + 1$ are the result of multiplying all neuron values in layer i by their individual weights and passing the result to the activation function. In the last layer, a single neuron value is calculated and becomes the regression Y' for the objective variable. The square error $(Y - Y')^2$ of this Y' and the explanatory variable Y is entered in the cell to the right of Y'.

In the third and subsequent rows, the sheet function formulas entered in the two rows are copied for (number of observations -1) rows. In a cell one row apart from the last row of all copied data, enter the formulas for the mean square error and the mean square error. The weight labels and values are entered in the two columns to the right. For clarity, the columns with the neuron labels and values corresponding to each layer are colored.

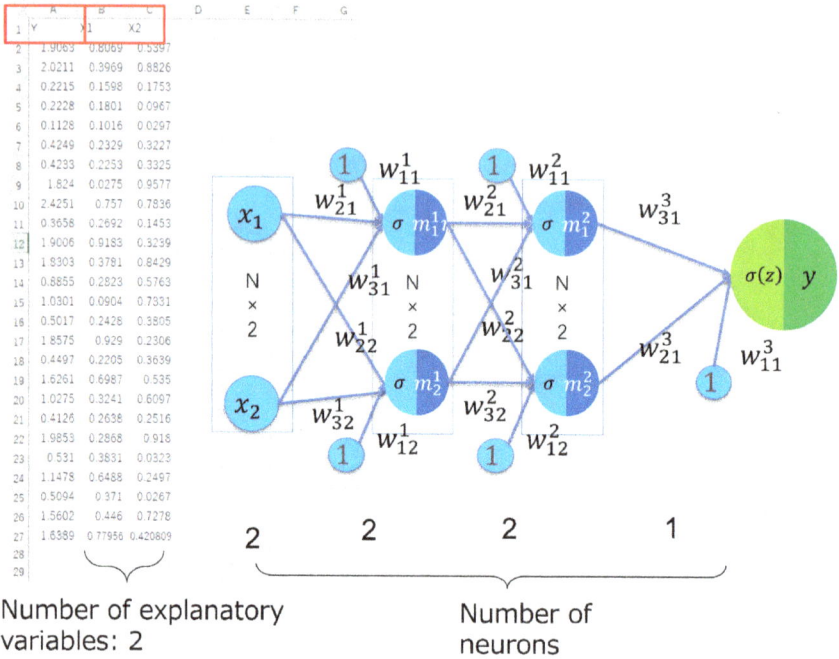

Figure 2.31. Creation of NN (hidden layer 2) model.

The weights are also colored accordingly for each layer. The Excel Solver obtains the weights that minimize the mean squared error. This completes the NN model.

Preparation for programming with Excel VBA includes the execution of the Solver tool and obtaining macro output for row copying. Turn on the Record Macro, execute the Solver tool, and copy certain rows. In some cases, the macro for Solver tool execution may not work properly. This can be resolved by performing the following configuration.

(1) *Check the reference settings for the Solver*: Make sure the Solver used by the macro is referenced to the correct location. In Visual Basic Editor, click [Tools] — [References] and make sure the Solver is checked.

(2) *Check the security settings*: Make sure that the security settings required to run macros are enabled. In Excel Options, click [Trust Center] — [Macro Settings], and select [Enable VBA macros].

Excel Solver can be combined with macros to automatically perform optimizations or solve large numbers of problems at once. When using

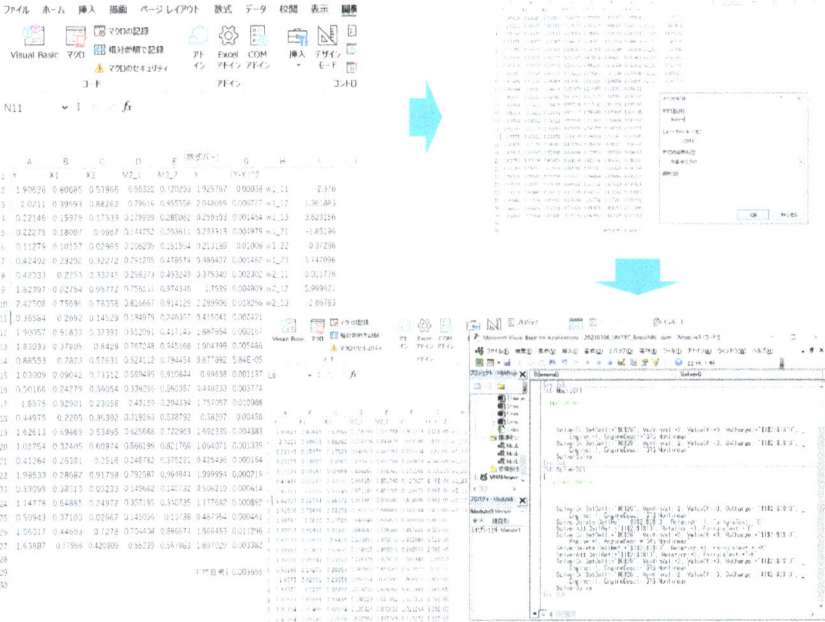

Figure 2.32. Creation of a macro which includes a solver.

macros, the objective function, constraints, range of variables, and direction of constraints must be set in VBA (Figure 2.32).

The following is an example of a Solver using VBA:

```
<Begin>
Sub SolveWithSolver()
    Dim SolverOK As Boolean
    SolverOK = Application.AddIns("Solver Add-in").Installed
    If Not SolverOK Then
        MsgBox "Solver add-in not installed"
        Exit Sub
    End If
    Range("A1").Select
    SolverReset
    SolverOk
    SetCell:="$B$4", MaxMinVal:=2, ValueOf:=0,
        ByChange:="$B$2:$B$3", Engine:=1,
        EngineDesc:="GRG Nonlinear"
```

```
SolverAdd CellRef:="$B$2", Relation:=1, FormulaText:="> =0"
SolverAdd CellRef:="$B$2", Relation:=3, FormulaText:="< =50"
SolverAdd CellRef:="$B$3", Relation:=1, FormulaText:="> =0"
SolverAdd CellRef:="$B$3", Relation:=3, FormulaText:="< =50"
SolverSolve UserFinish:=True
End Sub
<End>
```

In this code, the explanatory variables are set to maximize B4 within the range of B2 and B3. The SolverAdd method is used to set range constraints on B2 and B3. Lastly, the SolverSolve method is used to find the best solution. In this way, VBA can be used to automate the Solver.

The next method is to obtain the macro output of the row copy. The following is an example of an Excel macro that copies the second row, columns 3 through 6, to row 3 using **Selection.AutoFill**:

```
<Begin>
Sub CopyRangeToColumn()
'Specify the range from which to copy
Range_desc1 = "C2:F2"
Range(Range_desc1).Select
'Specify the range to copy to
Range_desc2 = " "C3:F3" "
Selection.AutoFill Destination:=Range(Range_desc2),
Type:=xlFillDefault
End Sub
<End>
```

The above macro copies the values from columns C to F in the second and third rows. First, use the **Range()** method to specify the range to copy to. Then, use **Selection.AutoFill** to specify the copy source range. For **Destination**, specify the copy destination range. As described above, **Selection.AutoFill** can be used to automatically copy values from the start column to the end column.

When you have finished the preparation, in the first row, enter the objective variable, explanatory variables, and neuron variable labels as character-type data in the cells. For the objective variable, the variable in which the column number of the cell is stored is represented as Col_Var. The VBA program statement is as follows:

Cells(1, Col_Var) = "Y"

For the objective variable, Col_Var = 1. The second and subsequent columns will be the labels, Xj, of the explanatory variable, thus the VBA program statement is as follows:

```
<Begin>
    For j = 1 To neuron_num(1)
        Var_Label = "X" & CStr(j)
        Cells(1, Col_Var) = Var_Label
        Col_Var = Col_Var + 1
    Next j
<End>
```

In this program, neuron_num(1) contains the number of neurons (in this case, explanatory variables) placed in the first layer. In the case of neuron variables, the label name is Mi_j. Thus, the program statement is as follows:

```
<Begin>
    For j = 1 To neuron_num(i)
        Var_Label = "M" & CStr(i) & "_" & CStr(j)
        Cells(1, Col_Var) = Var_Label
        Col_Var = Col_Var + 1
    Next j
<End>
```

After the neuron labels for layer i are described, the process of supplying each neuron variable in the next $i + 1$ layer with the sum of products calculated by the Excel sheet function formula MMULT and the subsequent conversion results from the activation function is executed.

The weights for each neuron variable are then stored in the rightmost column of the sheet in layer order. The number of columns in which to store the weights is calculated. The number of neurons per layer is added in all layers, and the columns storing the squared error and the labels of the weights are arranged side by side, with the column storing the values of the weights to the right of them. The program statement to calculate the column number and obtain the column name of the corresponding column is as follows:

```
<Begin>
Col_W =1
For i = 1 To layer_num + 2
    Col_W = Col_W + neuron_num(i)
Next i
Col_W = Col_W + 3
Col_W_Char = Split(Cells(1, Col_W).Address, "$")(1)
<End>
```

In this program, Split(... obtains the address of the cell at the column number specified by **Cells(1, Col_W)**, splits it into an array separated by **"$"**, and stores its second element (i.e., the column name) in Col_W_Char.

The weights are defined and named for all fully connected arcs with neurons in the previous layer for each layer. The intercept is also included in the weights. Thus, it is only 1 larger than the number of neurons in the previous layer. Each neuron in layer i connects to all neurons +1 (intercept) in layer $i-1$. As described earlier, Col_W represents the row number on which the weights are described. W_desc is character-type data representing the label of the weights.

```
<Begin>
Row_W = 2
For i = 2 To layer_num + 2
  For j = 1 To neuron_num(i)
    For k = 1 To neuron_num(i - 1) + 1
        W_desc = "w" & CStr(i - 1) & "_" & CStr(j) & CStr(k)
        Cells(RowV, Col_W - 1) = W_desc
        Cells(RowV, Col_W) = Rnd()
        Row_W = Row_W + 1
    Next k
  Next j
Next i
<End>
```

Next, we calculate the column number pairs (Col1, Col2: both character types) in which the variable values subject to MMULTed are stored. Only the second row is calculated, thus it is concatenated with a 2 at the end. After that, it copies as many observations as there are.

```
<Begin>
Col1 = Split(Cells(1, Target_Col).Address, "$")(1)
Col1 = Col1 + CStr(2)
Col2 = Split(Cells(1, Target_Col - neuron_num(i) + 1).Address, "$")(1)
Col2 = Col2 + CStr(2)
<End>
```

A detailed explanation follows: **Cells(1, Target_Col)** represent the cell with the column number specified by **Target_Col** in the first row. The **Address** property is used to obtain the address of this cell.

Cells(1, Target_Col).Address returns a string, such as **$B$1** for column **B**.

Next, the string is split into arrays, separated by **"$"**. For example, **Split(Cells(1, Target_Col).Address, "$")** returns an array like this for column **B**:

$$["", "B", "1"]$$

The objective of this code is to obtain the column name of the column corresponding to the specified column number (Target_Col). By obtaining the column name, you can specify a cell from the column name and row number, for example, Range(Col1 & "2").

2.7.4. *Function approximation using NNs*

To explain the function approximation using NNs, we use as discrete data the bivariate data described in Section 5, Paragraph 2. To use ExNN, first assume that the Excel sheet, the objective variable data is populated in the first column, starting from the second row and continuing for as many observations as there are, then, assume that the explanatory variable data is populated in the second and subsequent columns, starting from the second row, with a number of explanatory variables equal to the number of observations. For the structure of the NN used in the calculation, we consider the following six structures, all with 2 hidden layers. The numbers represent the number of neurons placed in each layer. The number of layers is assumed to increase sequentially from left to right, starting from 1.

Structure 1: 2,2,2,2,1
Structure 2: 2,3,3,3,1
Structure 3: 2,1,2,3,1
Structure 4: 2,2,3,4,1
Structure 5: 2,3,2,1,1
Structure 6: 2,4,3,2,1
Structure 7: 2,3,2,3,1
Structure 8: 2,4,3,4,1
Structure 9: 2,2,3,2,1
Structure 10: 2,3,4,3,1

Structures 1 and 2 represent a pattern in which the number of neurons is constant (constant NN), structures 3 and 4 represent an increase in the number of neurons (increasing NN), structures 5 and 6 represent a decrease in the number of neurons (decreasing NN), structures 7 and 8 represent the number of neurons once decreasing and then increasing (concave NN), and structures 9 and 10 represent the number of neurons once increasing and then decreasing (convex NN). The first neuron number represents the number of input variables (2 in this case), and the last neuron number represents the number of output variables (1 in this case). Figure 2.33 shows an example of a convex BB structure.

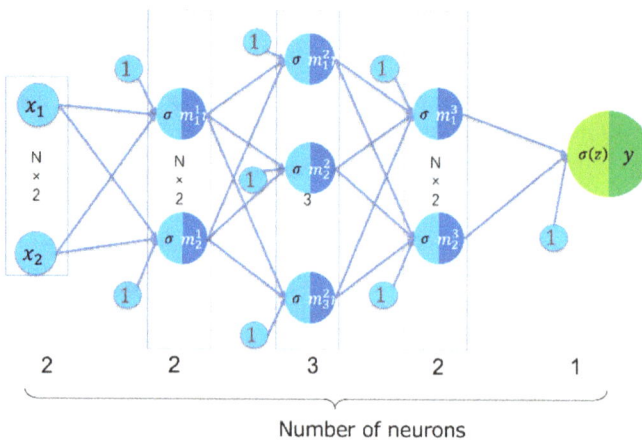

Figure 2.33. NN (Intermediate layer 3) structure; convex.

For each structure, we use the bivariate data described in Section 5, Paragraph 2 as discrete data to evaluate the model error of the NN. These data are calculated from the bivariate function

$$y = x_1^2 + x_1 + x_2^3 + x_2^2. \tag{2.32}$$

The results show that the model error with convex NNs is the smallest.

In NNs, the number of neurons per layer is variable. For example, an NN can have a structure that gradually reduces (decreasing type), increases (increasing type), or keeps constant (constant type) the number of neurons from the input layer. There are also other patterns (convex, concave, etc.). As a visualization method for naturally representing such parameters, NNs are represented by setting an upper limit on the number of layers (dimensions) and using the number of neurons per layer as a multidimensional coordinate point. One way to visualize multidimensional coordinates is to use parallel coordinates. Parallel coordinates assign each dimension of the multidimensional coordinate to a parallel coordinate axis and represent it as a polyclonal line connecting the values for each dimension. Figure 2.34 represents six types of NNs, each consisting of four layers, using parallel coordinates. The intersection in each coordinate axis represents the number of neurons in that layer. The rightmost coordinate axis represents the error.

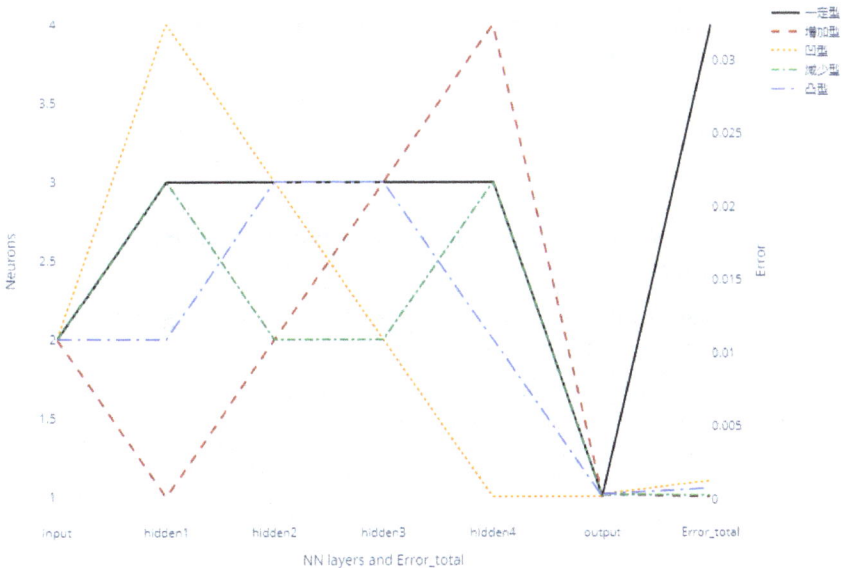

Figure 2.34. NN model error evaluation (ExNN).

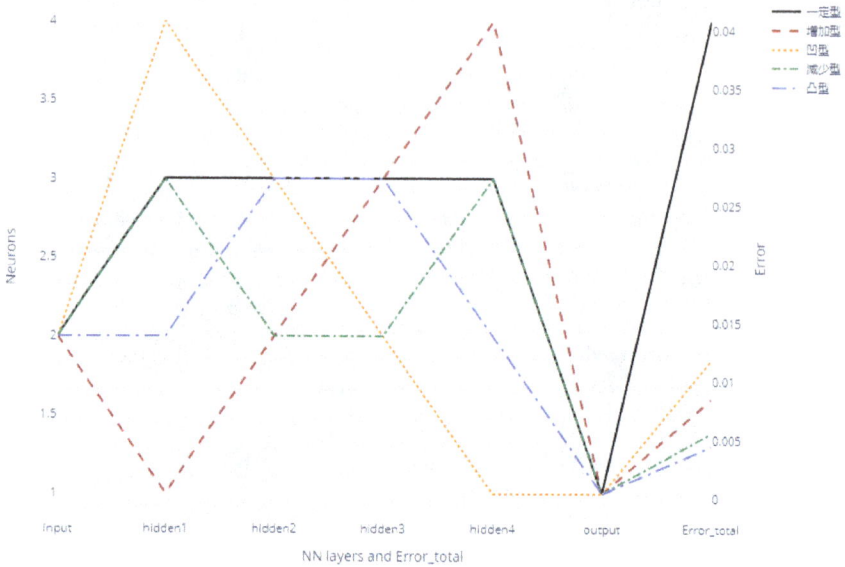

Figure 2.35. NN model error evaluation (keras).

This polygonal line is the coordinate point in the multidimensional space. In this example, we note that the error in the convex NN is smaller.

NN models of the same type of structure are created from the same physical data using Excel NN and keras.

In both cases, the error is the largest in a constant type NN. ExNN has a small error in increasing NN, and keras has a small error in convex NN (Figure 2.35). In addition, the error of ExNN tends to be small in general.

2.7.5. *Point cloud data analysis using NNs: Application example*

Assuming that physical data are interpreted as scalar data associated with coordinates, NN can be used to generate continuous data based on discrete data obtained from measurements. For the continuously distributed scalar data, various visualization methods can be applied. Furthermore, such scalar data can be used to develop new visualization methods for three-dimensional images.

For example, three-dimensional images captured by CT and other methods play an important role in non-invasive, internal structural analysis, as well as in preoperative diagnosis in the medical field. Specifically, the scalar data can be used to define new surfaces for mapping three-dimensional images.

Assuming that a three-dimensional image of a machine part has been obtained, the areas that can be determined to be the points on the crack surface can be identified by mouse operation, etc., and the data stored in the system along with their certainty value (scalar value). NN can be trained on the stored scalar data (physical data) to build a continuously distributed confidence model. An ordered lattice is defined in the same region as the three-dimensional image, and data are retrieved from the confidence model at the lattice points to generate the ordered lattice data.

When visualizing the page surfaces of ancient documents that cannot be turned, the page planes are extracted, and the three-dimensional images of the documents are mapped on the page planes. In this case, the area that can be determined as a point on the page surface is identified by a mouse operation, etc., and stored in the system along with its page number (scalar value) (Figure 2.36).

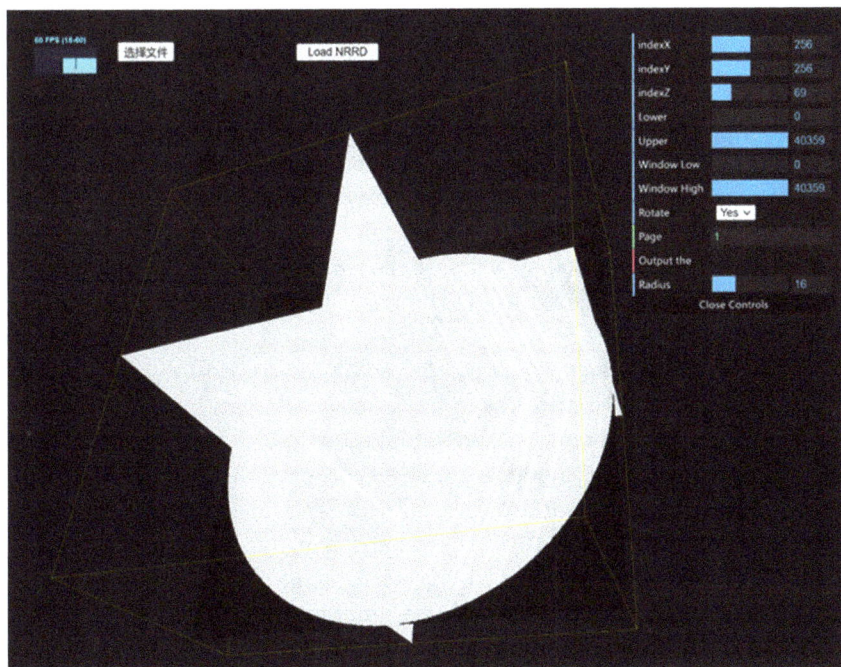

Figure 2.36. Page surface identification by mouse operation.

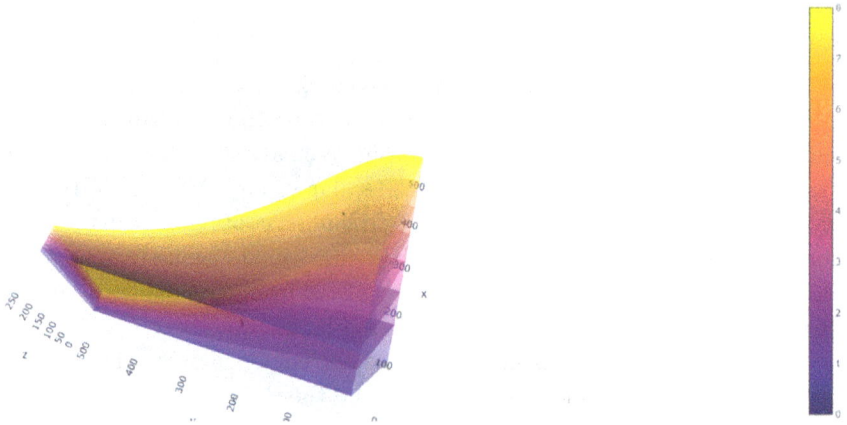

Figure 2.37. Page surface visualization using isosurfaces.

NN can be trained on stored scalar data (physical data) to build a continuously distributed page number model. For example, an isosurface for a scalar value of 2 represents the page surface of page 2 (Figure 2.37).

There are various methods for visualizing ordered lattice data. In this section, we use the Marching Cubes method, which can effectively visualize isosurfaces. The Marching Cubes method approximates the surface of an object in 3D space using scalar data defined on a 3D lattice. Specifically, the intersection points with isosurfaces are calculated on the ridges of the 3D lattice. Using these points, a triangular mesh that approximates the isosurface is calculated to approximate the surface of an object in 3D space. This method is widely used in 3D computer graphics and visualization of 3D scan data.

Next, the texture coordinates of the 3D image are interpolated and calculated at each vertex of the triangle mesh. Texture coordinates comprise a coordinate system used in computer graphics, such as 3D graphics and game engines, to attach images or patterns to the surface of objects. Normally, the surface of an object is divided into small shapes, such as triangles, and texture coordinates are assigned to each shape. Texture coordinates are defined for each shape's vertex and are usually expressed in two-dimensional coordinates. This allows for repeated placement and deformation of images or patterns on the surface of an object. Texture coordinates are used by a technique called texture mapping. Texture mapping can greatly enhance the expressive power of 3D

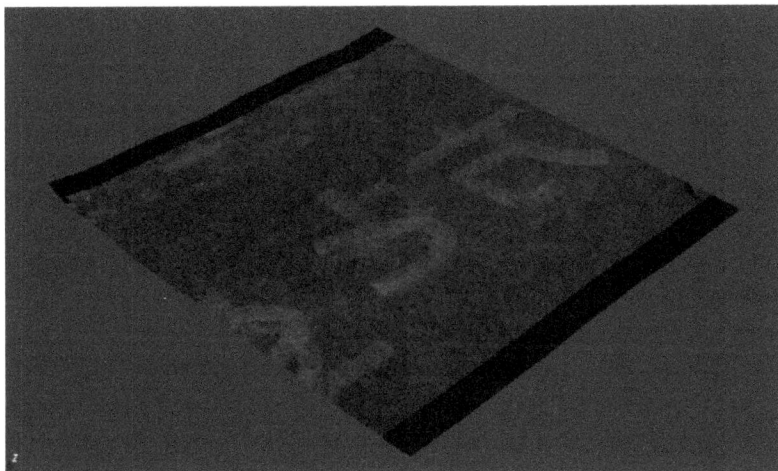

Figure 2.38. Page information visualization using texture mapping.

graphics by providing objects with realistic textures and details. When mapping three-dimensional images, such as ancient documents, the texture coordinates are three dimensional (Figure 2.38).

2.7.6. *Line group data analysis using NN application examples*

We have discussed the visualization of point cloud data using NN as an application example. Next, we would like to consider line group data as a visualization target. For example, when an appropriate starting point is placed in the flow field obtained from a fluid analysis, and massless particles are scattered along the flow field from that point, the curve created by the trajectory is called a streamline. By setting many starting points in the flow field, a group of streamlines can be generated. Sometimes, a display of the enveloping surface that encompasses this group of streamlines is required. The display of the enveloping surface is required not only for velocity vectors in fluid analysis but also for principal stress vectors in structural analysis, magnetic flux density vectors in electromagnetic field analysis, and so on.

This section describes a method in which NN is trained with magnetic field group data extracted from electromagnetic field analysis results calculated in a fusion reactor and calculates the probability of a plasma region as volume data. This method is an important technique for visualizing the plasma region in a fusion reactor. As mentioned above, vector line groups such as streamlines calculated from fluid analysis results can also be targeted.

A fusion reactor is a device that uses a nuclear fusion reaction to generate energy. Fusion is a reaction in which nuclei coalesce to form new nuclei and occurs at high temperatures and pressures. By creating such conditions in a fusion reactor, the hydrogen nuclei coalesce to produce a new nucleus of hydrogen — helium. In this process, a large amount of energy is generated. Fusion reactors have the potential to provide energy on Earth in the long term because they can generate energy without the use of petrochemicals such as oil or natural gas. However, in reality, constructing a fusion reactor is very difficult and has not been commercialized to date.

The structure of a fusion reactor consists of a vessel to generate and hold plasma, a magnetic field generator to control the plasma, a cooling system to extract heat from the plasma, and a fuel injection system to promote the fusion reaction.

The major components are as follows:

1. *Vessel*: A vessel that holds the plasma is located at the center of the fusion reactor. This vessel is placed in a vacuum and is made of very tough materials to hold high temperatures and high-density plasma. Typical materials for the vessel include tungsten alloys and silicon carbide.
2. *Magnetic field generator*: To control the plasma, a magnetic field must be generated. In fusion reactors, a tokamak/helical magnetic field arrangement is mainly used. In a tokamak/helical magnetic field arrangement, the force of the magnetic field keeps the plasma in a doughnut shape. The magnetic field is generated by superconducting coils.
3. *Cooling system*: To control the hot plasma produced by the fusion reaction, heat must be extracted. The cooling system is installed outside the plasma vessel and circulates coolants such as water or helium to extract the heat.
4. *Fuel injection system*: In a fusion reactor, light hydrogen (hydrogen with a single nucleus) is used as fuel. The fuel is injected into the plasma in a vacuum. Fuel injection systems include toroidal pumps and gas injectors.
5. *Divertor*: A divertor is a device used to separate and remove some of the protons and ions that are injected from outside the plasma. Divertors play an important role in stabilizing the operation of a fusion reactor by extracting heat and particles from the high-temperature, high-density region of the plasma.

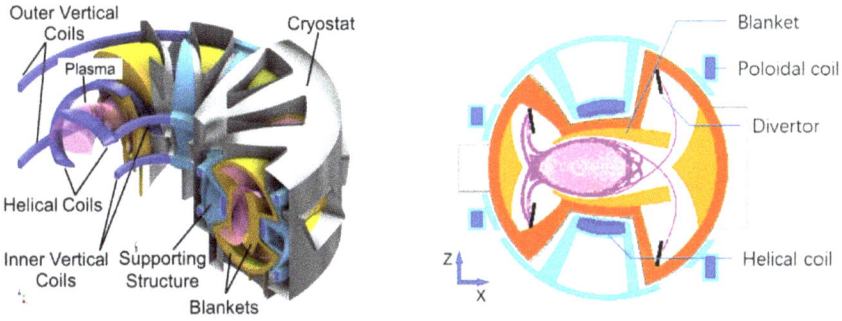

Figure 2.39. Structure of helical fusion reactor.

These are the main components of a fusion reactor. Figure 2.39 shows the structure of a helical fusion reactor.

In a fusion reactor, an electromagnetic field is used to confine and maintain the plasma (high-temperature gas) inside a container. The container housing the plasma is called a "tokamak" or "helical". Electromagnetic fields refer to electric and magnetic fields in terms of electromagnetic dynamics. Electromagnetic fields reflect the electrical behavior of charged particles and the magnetic behavior of magnetic materials. In a fusion reactor, electromagnetic coils can be used to create an electromagnetic field to confine the plasma. This electromagnetic field exists within the container that confines the plasma and serves to anchor it. In addition, in a fusion reactor, the temperature inside the plasma must be adjusted so that the fusion reaction does not occur inside the plasma. Therefore, electromagnetic field simulations are used in fusion reactors to study how to confine the plasma and regulate the temperature in the plasma. Next, we will discuss the edge and divertor, which play important roles in the structure of a fusion reactor.

In a fusion reactor, the edge is the physical boundary between the plasma and the wall. The plasma is composed of energetic particles produced at high temperatures and pressures, and the walls are typically made of cold metal. Therefore, there is a difference in temperature and pressure between the plasma and the wall. The boundary where this difference occurs is called the "edge". The edge plays an important role in fusion reactors. Among the most important roles are the following:

- It serves to separate the plasma within the container that houses the plasma. This prevents the plasma from damaging the walls.

- It serves to prevent the loss of energy within the plasma. If energy is lost in the plasma, the fusion reaction slows down, and energy cannot be produced. It prevents energy from being lost in the plasma through the edge.

The edge plays a critical role in a fusion reactor. Therefore, it is important to constantly monitor and maintain the condition of the edge.

In a fusion reactor, a divertor refers to a device that separates the plasma. Plasma, as explained earlier, consists of high-energy particles produced at high temperatures and pressures. As a result, the plasma can damage walls. Therefore, it is necessary to separate the plasma. A divertor is a device that separates such plasma by changing the direction in which the plasma flows. The divertor is usually installed inside the container that contains the plasma. By changing the plasma's flow direction, the plasma can be separated. Divertors play an important role in fusion reactors. Among the most important roles are the following:

- It serves to separate the plasma within the container that houses the plasma. This prevents the plasma from damaging the walls.
- It serves to prevent the loss of energy within the plasma. If energy is lost in the plasma, the fusion reaction slows down, and energy cannot be produced. Through the divertor, energy can be prevented from being lost in the plasma.

The divertor plays a critical role in a fusion reactor. Therefore, it is important to constantly monitor and maintain the condition of the divertor.

In a fusion reactor, it is expected to visualize the plasma region from a group of magnetic field lines calculated by setting an appropriate starting point based on the electromagnetic field simulation results. A device called the Force Free Helical Reactor (FFHR), designed by the National Institute for Fusion Science (NIFS), has helical coils to generate twisted magnetic field lines [51]. The plasma is confined within a donut-shaped vacuum vessel. The large-circumferential direction of the donut-shaped vacuum vessel is called the toroidal direction, and the small-circumferential direction is called the poloidal direction. In the helical type, as the name implies, an electric current is applied to the helical circumferential coils to form a magnetic field that confines the plasma, and the enveloping surface of the magnetic flux is considered the boundary of the plasma region.

One of the main problems with the device designed by FFHR is that the space between the plasma region geometry and the structure in the vessel is limited. Conventional visualization of the plasma region mainly calculated the intersection of magnetic field lines in the poloidal plane of the plasma and expressed the plasma region on a two-dimensional cross section. Based on this visualization, the geometry of the in-vessel structures was examined to determine whether they were in contact with the plasma boundary. This method was performed by rotating the poloidal surface one degree in the toroidal direction, which was not an efficient way to get a bird's-eye view of the plasma region shape. Therefore, in order to more efficiently determine the contact between the plasma and the structure, we considered it necessary to construct a 3D model of the plasma shape to determine the contact with the in-vessel structure. The objective of this case study is to create a 3D model representing the plasma existence region as a closed curved surface based on the magnetic field group data in order to facilitate identification of interference between the plasma and the 3D design data of the in-vessel structure components.

In this section, we will present an overview of electromagnetic field simulations in a fusion reactor. To calculate the steady-state magnetic field lines, 3D modeling of the internal environment of the FFHR is performed by inputting the precise structure of the magnetic field coils, vacuum vessel, and blanket and then calculating the electromagnetic field simulation using the finite element method (FEM), which is described later in 4.3.2.1. To trace the magnetic field lines from a suitable starting point, the magnetic field line trace code (MGTRC) is used. At present, 799 magnetic field lines are stored. And all magnetic field lines consist of 4.8×10^6 points (Figure 2.40).

As shown in Figure 2.40, the calculated 799 magnetic field lines are twisted, making it difficult to assess their intersection with the structure based on the magnetic field lines alone. The key is to detect interference with components in the fusion reactor, which requires evaluation of the outermost magnetic field lines.

90 poloidal surfaces with intersections with magnetic field lines are set up so the plasma area shape is 90° in the toroidal direction. Additional starting points are needed to draw the magnetic field lines leading to the surface layer of the edge and the legs of the diverter. A divertor is one of the devices that make up a fusion reactor and is responsible for three functions: particle pumping, heat removal, and improved plasma confinement. These starting points are indicated interactively using the Point Annotation Tool [52]. This process should be completed before the plasma region geometry is expanded to 90°.

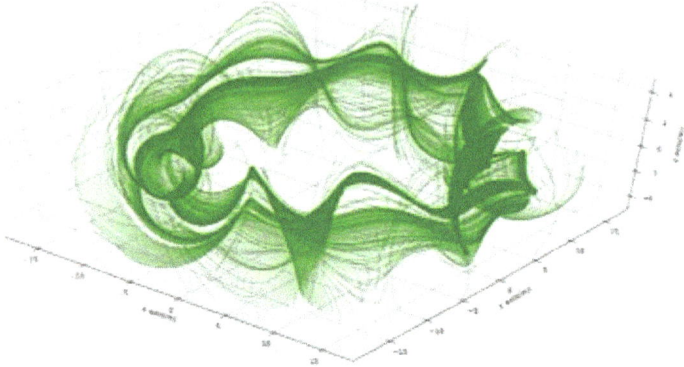

Figure 2.40. Group of magnetic field lines calculated in the structure of a helical fusion reactor.

To find the outermost closed curved surfaces of the magnetic and edge layers, labeling is required to indicate the trajectory of the points on the Poincaré plot. If the magnetic field lines stay in the plasma region long enough, all the points that make up the magnetic field lines are labeled with a "2". At the same time, all points comprising the magnetic field lines reaching the outer wall of the fusion reactor or the divertor are labeled "0" or "1", respectively. Magnetic field lines that reach the outer wall of the fusion reactor or the divertor are considered to have escaped outside the plasma region.

In this case study, a suitable NN model is first created that explains the point data on the magnetic field lines (Figure 2.40). NN model building is an NN parameter optimization calculation that minimizes the loss function. The loss function is the mean square error $\text{MSE}_u = \|\hat{u} - u\|_2^2 / N_u$ between the point data $u(x, y, z)$ and the data regressed by the NN model, where N_u represents the number of points on the magnetic field lines. In this case, the point data represent the probability that the point is in the plasma region and take a value of 1.0 on the magnetic field lines that stay within the plasma region. The information for the NN to perform this calculation is as follows:

- *Structure*: The NN has a five-layer structure with a number of nodes of 3-512-128-16-3 in the forward direction. The first three and the last three are the number of nodes in the input and output layers, respectively.

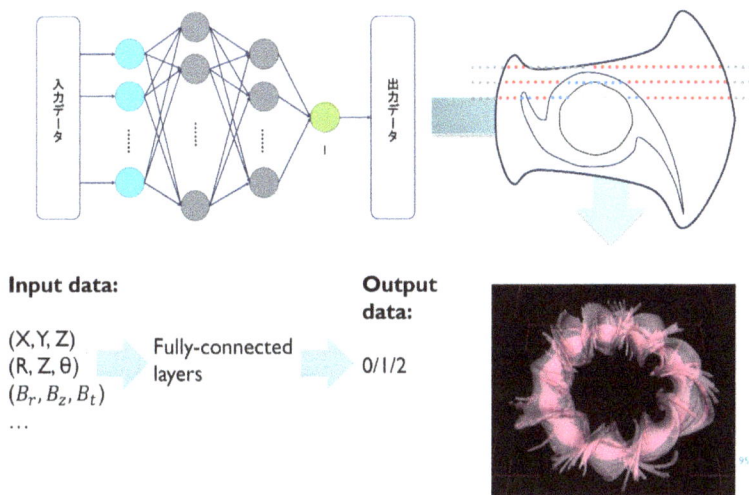

Figure 2.41. Visualization of the envelope surface of a group of magnetic field lines using isosurfaces.

- *Activation function*: Rectified linear unit activation function
- *Optimization calculation*: RMSProp

Figure 2.41 presents a visualization of the envelope surface formed by a group of magnetic field lines passing through a region identified as a plasma region and a divertor, represented as an isosurface.

Chapter 3

Practical Part

So far, we have explained how to create a spatiotemporal model using an NN from physical data. In terms of application examples, we have explained the method used for visualizing page information from ancient literature data shot utilizing a three-dimensional CT machine. In addition, we have explained the method for visualizing the plasma region from a magnetic line group calculated from the results of analyzing electromagnetic fields in a fusion reactor.

In this chapter, we will begin by explaining differentiation operations on a spatiotemporal model. The ability to perform differentiation operations may improve the accuracy of an NN model in some cases, and also enhance the accuracy with which to evaluate gradient data, which is important for visualization. It allows you to find an essential partial differential term and derive a PDE governing the given physical data. You can also solve the given PDE. As described earlier, adding differential information increases the added value of the NN model. Since differential information is crucial for a PDE that describes physics, we refer to such NN as physics-informed neural network (PINN).

3.1. About PINNs

An NN suitable for analyzing physical data is a physics-informed neural network (PINN). A PINN is a kind of machine learning model proposed by Raissi *et al.* that incorporates the laws of physics into the structure of an NN itself [28]. This allows a model to make predictions that are consistent with the underlying physics based on which a system is modeled. A PINN is particularly useful for solving problems, such as fluid dynamics, structural dynamics, and electrodynamics problems, where the relationship between variables is governed by the laws of physics. Incorporating these laws into the model allows the PINN to make more accurate predictions than a conventional NN that does not consider physical constraints. The PINN is trained using a combination of supervised and unsupervised learning. The supervised learning is used to adapt the model to training data, while the unsupervised learning is used to apply physical constraints. This allows the model to learn from both data and the underlying laws of physics.

In the PINN, the training data are spatiotemporal coordinates and physical data. The spatiotemporal coordinates are fed into the input layer. The weight parameters are optimized so that the NN results fit the physical data. This optimization represents supervised learning (Figure 3.1).

Next, a PDE is constructed that represents the laws of physics from the NN model by differential operations. Weight parameters that are not optimized produce PDE residuals (Figure 3.2).

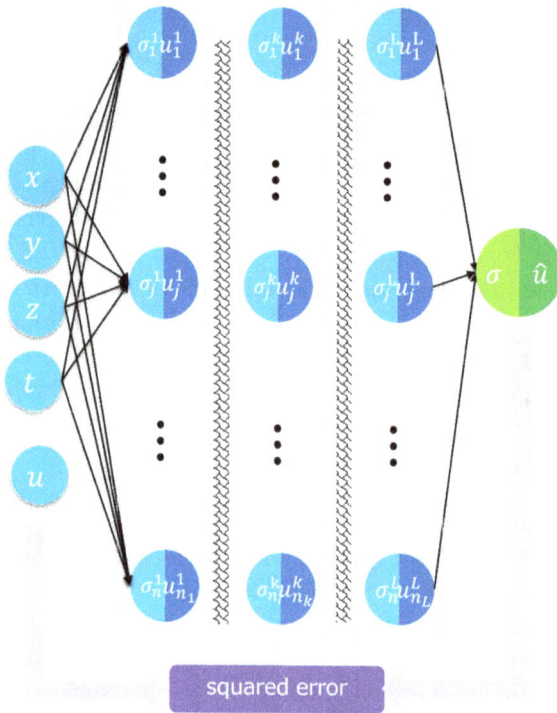

Figure 3.1. Supervised learning in PINNs.

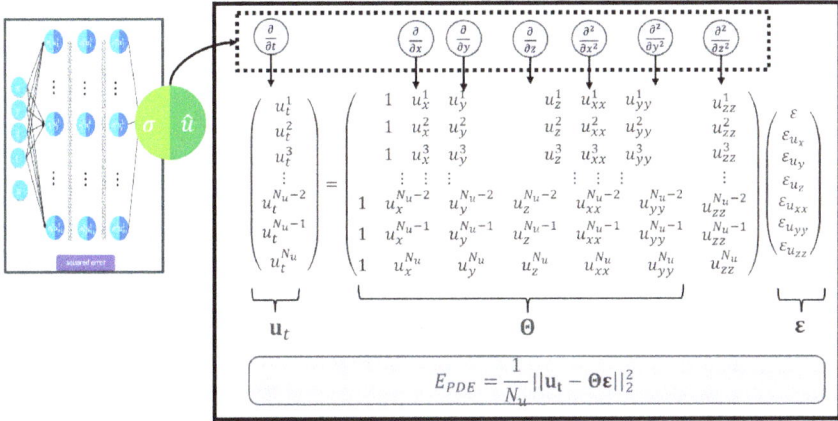

Figure 3.2. PDE residuals in PINNs.

To solve a PDE, a boundary condition is needed to fully determine the problem. The boundary condition provides information about the behavior of the system at the boundary of an analysis domain. In general, the analysis domain of the PDE has a boundary. For example, in the case of a two-dimensional PDE, the analysis domain is usually a plain with its boundary in the form of a line, circle, curve, and so on. This boundary must have the following boundary condition:

1. *Dirichlet condition*: Specifies that the value of a solution is known on the boundary. For a temperature equation, for example, the Dirichlet condition specifies that a wall or surface temperature is known (Figure 3.3).
2. *Neumann condition*: Specifies that the normal derivative of a solution is known on the boundary. For the heat conduction equation, for example, the Neumann condition specifies that heat flux is known on a wall or surface (Figure 3.4).

The boundary condition, which is needed to solve a PDE, is always handled as part of the problem. It is important to set the boundary condition properly, failing which, one may not obtain a correct solution.

The boundary condition required to solve a PDE is also constructed from the NN model, resulting in residuals. The Neumann condition requires a differential operation. The optimization of weight parameters is performed with the aim of zeroing out these residuals and represents unsupervised learning. The PINN calculates weight parameters that minimize

$$u\Big|_s = u(t^{BC}, x^{BC}, y^{BC}, z^{BC}) = u_0$$

$$u(t, x, y, z)$$

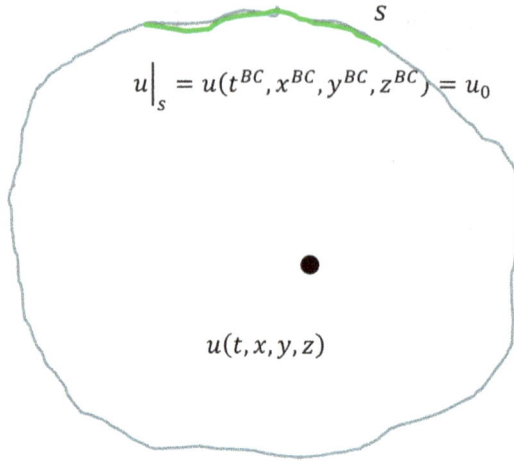

Figure 3.3. Dirichlet condition in PDE.

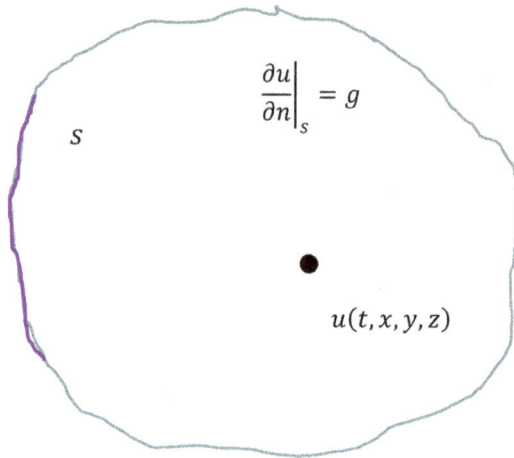

$$\frac{\partial u}{\partial n}\Big|_s = g$$

$$S$$

$$u(t, x, y, z)$$

Figure 3.4. Neumann condition in PDE.

a value obtained by adding these residuals to the loss function defined in supervised learning.

Differentiation operations are required to construct a PDE that represents laws of physics from the PINN model and the Neumann boundary condition, so one of the important technologies that support the PINN is automatic differentiation. Automatic differentiation or algorithmic differentiation refers to a technique of analyzing a function defined in a program to derive a program that calculates a partial derivative value [24]. Even a complex program consists of combinations of basic arithmetic operations, such as addition, subtraction, multiplication, and division, and basic functions, such as exponential, logarithmic, and trigonometric functions. Automatic differentiation takes advantage of this feature to find a solution by repeatedly applying the chain rule to these arithmetic operations. Automatic differentiation allows you to obtain a partial derivative value automatically with less computational effort.

3.2. Automatic differentiation in NN

This section describes how to calculate a 1st-order partial derivative with respect to an input variable of an FCNN, which is widely used in deep learning [25, 26]. Figure 3.5 shows the structure of FCNN in PINN.

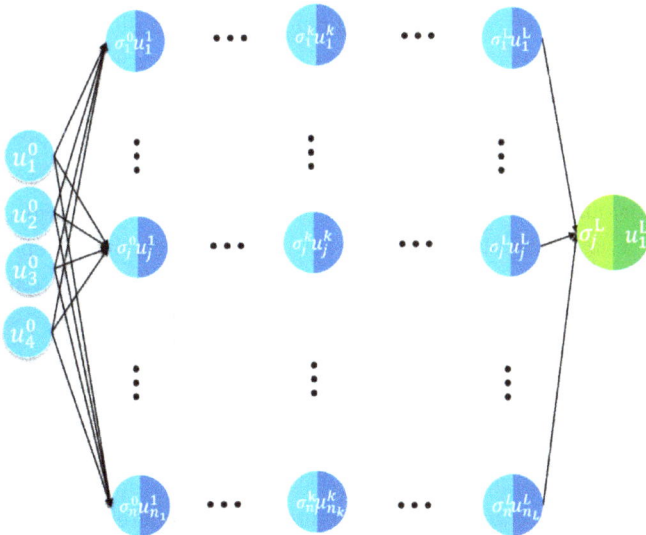

Figure 3.5. Structure of FCNN in PINN.

This NN is composed of an input layer and an output layer, with a hidden layer between them. The neurons in each layer are assumed to be fully connected to those in adjacent layers. The kth neuron j in the u_j^kth layer is connected to the neurons in the $(k-1)$th layer, expressed as:

$$u_j^k = \sigma_j^{k-1}\left(\sum_{i=1}^{n_{k-1}} w_{i,j}^{k-1} u_i^{k-1}\right) \tag{3.1}$$

where $w_{i,j}^{k-1}$ is a weight for units u_i^{k-1} to u_j^k, and σ_j^k and n_k represent an activation function for the jth neuron in the k layer and the number of neurons, respectively.

Let us assume that the sigmoid function $\sigma(x) = 1/(1+e^{-x})$ is used as the activation function. The calculation is performed with the weight set to $w_{n_k,j}^k$ by setting bias for the kth layer to $u_{n_k}^k = 1$. The following paragraphs describe how to calculate the 1st-order derivative $\frac{\partial u_1^{L+1}}{\partial u_j^0}$ for the input layer u_j^0 and for the output layer unit u_j^{L+1}.

In the 1st-order derivative calculation, the chain rule is applied to decompose $\frac{\partial u_1^{L+1}}{\partial u_j^0}$ to construct an iterative procedure. u_j^{k+1} is calculated using $u_i^0, u_i^1, \ldots, u_i^k$. Therefore, u_j^{k+1} can be regarded as a composite function of $u_i^0, u_i^1, \ldots, u_i^k$, and its partial derivative is calculated using the chain rule. A partial derivative with respect to the input variable is calculated by recursively applying the chain rule decomposition of the partial derivative to this composite function. In this procedure, $\frac{\partial u_1^{L+1}}{\partial u_j^0}$ is decomposed according to the chain rule for intermediate variable u_i^1, and its subexpressions are further decomposed according to the chain rule for the intermediate variable u_i^2. The following formula is obtained by recursively repeating this decomposition process until the intermediate variable $u_{m_L}^L$ is reached.

$$\frac{\partial u_j^{L+1}}{\partial u_i^0} = \sum_{m_1=1}^{n_1}\left(\sum_{m_2=1}^{n_2}\left(\sum_{m_3=1}^{n_3}\left(\cdots\left(\sum_{m_L=1}^{n_L}\frac{\partial u_j^{L+1}}{\partial u_{m_L}^L}\frac{\partial u_{m_L}^L}{\partial u_{m_{L-1}}^{L-1}}\right)\cdots\right)\right.\right.$$

$$\left.\left.\times\frac{\partial u_{m_3}^3}{\partial u_{m_2}^2}\right)\frac{\partial u_{m_2}^2}{\partial u_{m_1}^1}\right)\frac{\partial u_{m_1}^1}{\partial u_i^0} \tag{3.2}$$

Based on this equation, a procedure is constructed that calculates a partial differential by propagating the NN in reverse order.

For example, when the activation function is the sigmoid function, $\sigma(x) = 1/(1+e^{-x})$, the derivative can be expressed as $\sigma'(x) = (1-\sigma)\sigma$

using the sigmoid function itself. The derivative between consecutive layers is expressed as follows:

$$\frac{\partial u_{m_{k+1}}^{k+1}}{\partial u_{m_k}^k} = u_{m_{k+1}}^{k+1}(1 - u_{m_{k+1}}^{k+1})w_{m_k,m_{k+1}}^k \tag{3.3}$$

Then, $\frac{\partial u_j^{k+1}}{\partial u_i^{k-1}}$ is decomposed according to the chain rule for partial differentiation with respect to the variable u_m^k in the kth layer. That is, a partial differentiation calculation that leaps one layer is performed as follows (Figure 3.6):

$$\frac{\partial u_{m_{k+1}}^{k+1}}{\partial u_{m_{k-1}}^{k-1}} = \sum_{m_k=1}^{n_k} \frac{\partial u_{m_{k+1}}^{k+1}}{\partial u_{m_k}^k} \frac{\partial u_{m_k}^k}{\partial u_{m_{k-1}}^{k-1}} \tag{3.4}$$

where the subexpression in Equation (3.4) is calculated from Equation (3.3). Similarly, the derivative with respect to the $k-2$th layer variable $u_{m_{k-2}}^{k-2}$

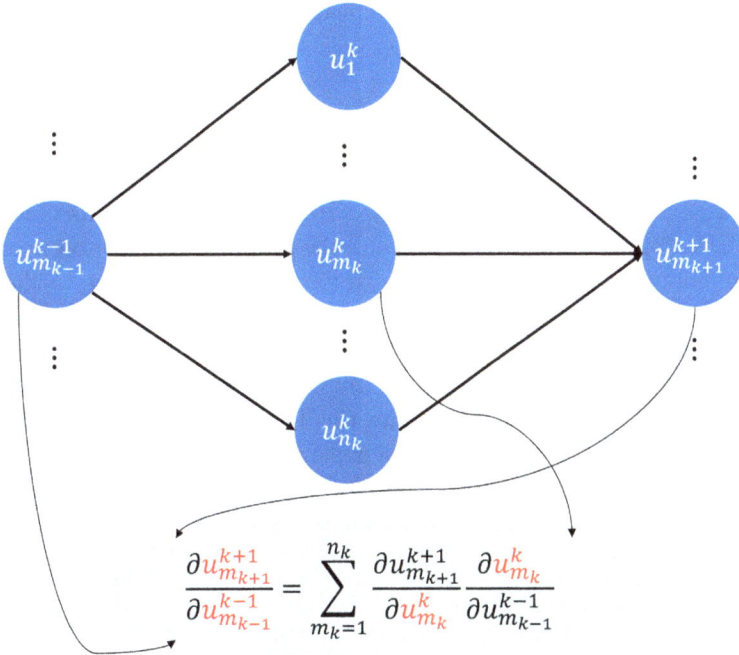

Figure 3.6. Calculation of partial differentiation according to chain rule in PINN.

is decomposed by the partial derivative with respect to the $k - 1$th layer variable $u_{m_{k-1}}^{k-1}$.

$$\frac{\partial u_j^{k+1}}{\partial u_{m_{k-2}}^{k-2}} = \sum_{m_{k-1}=1}^{n_{k-1}} \frac{\partial u_{m_{k+1}}^{k+1}}{\partial u_{m_{k-1}}^{k-1}} \frac{\partial u_{m_{k-1}}^{k-1}}{\partial u_{m_{k-2}}^{k-2}} \tag{3.5}$$

where the subexpression in Equation (3.5) is calculated from Equations (3.3) and (3.4). The differentiation can be performed with respect to the variables in the $k - n$ layer by repeating this process sequentially.

$$\frac{\partial u_{m_{k+1}}^{k+1}}{\partial u_{m_{k-n}}^{k-n}} = \sum_{m_{k+1-n}=1}^{n_{k+1-n}} \frac{\partial u_{m_{k+1}}^{k+1}}{\partial u_{m_{k+1-n}}^{k+1-n}} \frac{\partial u_{m_{k+1-n}}^{k+1-n}}{\partial u_{m_{k-n}}^{k-n}} \tag{3.6}$$

The subexpressions in Equation (3.6) are calculated from the partial differentiation for the mth and $k + 1 - n$th layers. $\frac{\partial u_1^{L+1}}{\partial u_j^0}$ is calculated by setting $k = L$ in Equation (3.6) and repeating the calculation by increasing n from 0 to L.

3.3. Using automatic differentiation in regression equation

We have shown regression analysis examples for a given bivariate function $(y = x_1^2 + x_1 + x_2^3 + x_2^2)$ using an NN. In section 2.7.2, we discussed the calculation of parameters that minimize the mean square error between the spatiotemporal model and the provided discrete data using the coefficients and NN weights as parameters. This calculation was performed using the Excel Solver function. By using a sigmoid function as an activation function, we successfully confirmed that the mean squared error (MSE) was reduced. This time, we will evaluate the 1st-order partial differentiation of a regression equation composed of explanatory variables (x_1, x_2) by two methods: The first method is to perform partial differentiation of a given function as it is, resulting in:

$$\begin{cases} \dfrac{\partial y}{\partial x_1} = 2x_1 + 1 & (3.7) \\[2mm] \dfrac{\partial y}{\partial x_2} = 3x_2^2 + 2x_2 & (3.8) \end{cases}$$

The second method is to find a partial derivative for the NN model obtained by regressing the given data. In this case, variables (m_1, m_2) in the hidden layers and the chain rule are used to calculate as follows:

$$\begin{cases} \dfrac{\partial y}{\partial x_1} = \dfrac{\partial y}{\partial m_1}\dfrac{\partial m_1}{\partial x_1} + \dfrac{\partial y}{\partial m_2}\dfrac{\partial m_2}{\partial x_1} & (3.9) \\[3mm] \dfrac{\partial y}{\partial x_2} = \dfrac{\partial y}{\partial m_1}\dfrac{\partial m_1}{\partial x_2} + \dfrac{\partial y}{\partial m_2}\dfrac{\partial m_2}{\partial x_2} & (3.10) \end{cases}$$

Now, using the sigmoid function $\sigma(x) = 1/(1 + e^{-x})$ as the activation function, each partial differential term on the right side of the above Equations (3.9) and (3.10) is expressed as follows:

$$\begin{cases} \dfrac{\partial y}{\partial m_1} = y(1-y)w_{11}^2 \qquad \dfrac{\partial y}{\partial m_2} = y(1-y)w_{21}^2 \\[3mm] \dfrac{\partial m_1}{\partial x_1} = m_1(1-m_1)w_{11}^1 \qquad \dfrac{\partial m_2}{\partial x_1} = m_2(1-m_2)w_{12}^1 \\[3mm] \dfrac{\partial m_1}{\partial x_2} = m_1(1-m_1)w_{21}^1 \qquad \dfrac{\partial m_2}{\partial x_2} = m_2(1-m_2)w_{22}^1 \end{cases}$$

That is,

$$\begin{cases} \dfrac{\partial y}{\partial x_1} = y(1-y)\left\{ w_{11}^2 m_1(1-m_1)w_{11}^1 + w_{21}^2 m_2(1-m_2)w_{12}^1 \right\} \\[3mm] \dfrac{\partial y}{\partial x_2} = y(1-y)\left\{ w_{11}^2 m_1(1-m_1)w_{21}^1 + w_{21}^2 m_2(1-m_2)w_{22}^1 \right\} \end{cases}$$

A mean squared error from the partial derivative calculated by these two methods is calculated as follows:

$$MSE_{grad} = \frac{1}{N_u}\sum_{i=1}^{N_u}\left\{ \left(\frac{\partial y}{\partial x_1} - (2x_1+1) \right)^2 + \left(\frac{\partial y}{\partial x_2} - (3x_2^2 + 2x_2) \right)^2 \right\}$$

$$(3.11)$$

The calculated mean squared error is added to MSE_u, the mean squared error that was minimized last time (Figure 3.7). The resulting sum was minimized with the Excel solver function, resulting in MSE_u being changed from 0.0081 to 0.00696. The reason for this improvement is as follows: For the partial differentiation of the regression equation of this time, the calculated result is already known, so we have successfully improved the predictive accuracy of the regression equation by using this information.

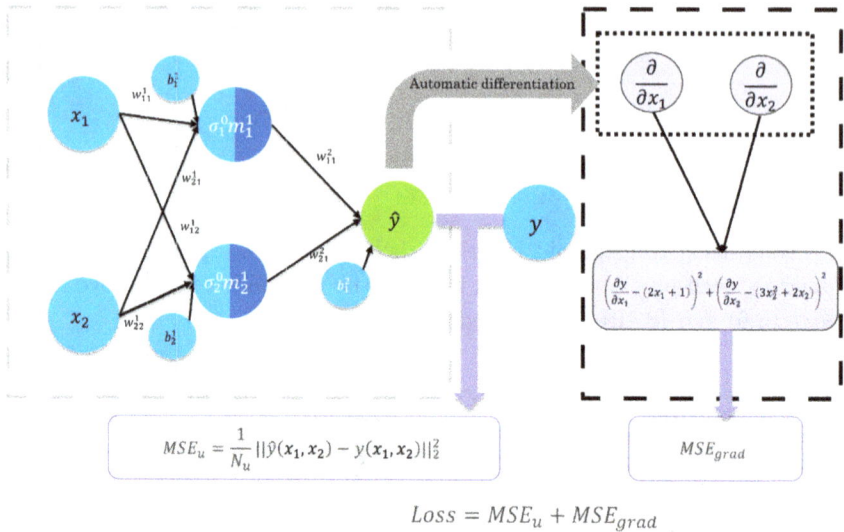

Figure 3.7. Adding gradient vector error to loss function.

3.4. Automatic differentiation using Colab

The automatic differentiation described in the previous section can be achieved using Colab. The following paragraph shows an example of an automatic differentiation program using Colab: TensorFlow is used here.

```
< Start >
import tensorflow as tf

# Defining input variables
x = tf.constant(3.0)
y = tf.constant(4.0)

# Defining a calculation chart
with tf.GradientTape() as tape:
  tape.watch(x)
  tape.watch(y)
  z = x ** 2 + 2 * x * y + y ** 2

# Calculating a gradient
grads = tape.gradient(z, [x, y])
```

```
print(grads)
```
< End >

The program defines **x** and **y** as constants and calculates **z** as a function of **x** and **y**. Then, **tf.GradientTape()** is used to define a calculation chart that calculates the partial derivatives of **x** and **y** with respect to **z**. **tape.gradient ()** calculates the partial derivatives of **x** and **y** with respect to **z**.

The execution result is displayed as follows: [<**tf.Tensor: shape=(),** **dtype=float32, numpy=10.0**>, <**tf.Tensor: shape=(), dtype=** **float32, numpy=10.0**>] This indicates that the partial derivatives of **x** and **y** with respect to **z** are each **10.0**.

3.5. Increasing the accuracy of visualization of magnetic line group data using automatic differentiation

As an example of analysis of line group data using an NN, we have explained the construction of an NN model that can describe point data on the magnetic lines calculated from simulation of the electromagnetic field in a fusion reactor. This model allows you to calculate the probability of an arbitrary point being in the plasma region. In calculating this probability $u(x, y, z)$, its gradient vector represents the normal vector to the plasma region surface. The gradient vector can be easily calculated by automatic differentiation. The magnetic flux density vector obtained from the electromagnetic field simulation results is parallel to the magnetic lines. So, the normal vector to the region surface and the magnetic flux density vector can be expected to be orthogonal. At each point on a magnetic line, inner products are calculated between the gradient vector and the magnetic flux density vector at that point, and then the inner products at all points on the magnetic line are added. Assuming that the NN model represents a plasma region, the added result is expected to be zero, so the following term is a new error term:

$$MSE_{ip} = \frac{1}{N_u} \left\| \left(\frac{\partial u}{\partial x} \quad \frac{\partial u}{\partial y} \quad \frac{\partial u}{\partial z} \right) \begin{pmatrix} B_x \\ B_y \\ B_z \end{pmatrix} \right\|_2^2 \qquad (3.12)$$

This error term is added to the loss function. When the loss function is small enough, the surface enveloping the magnetic lines can be visualized

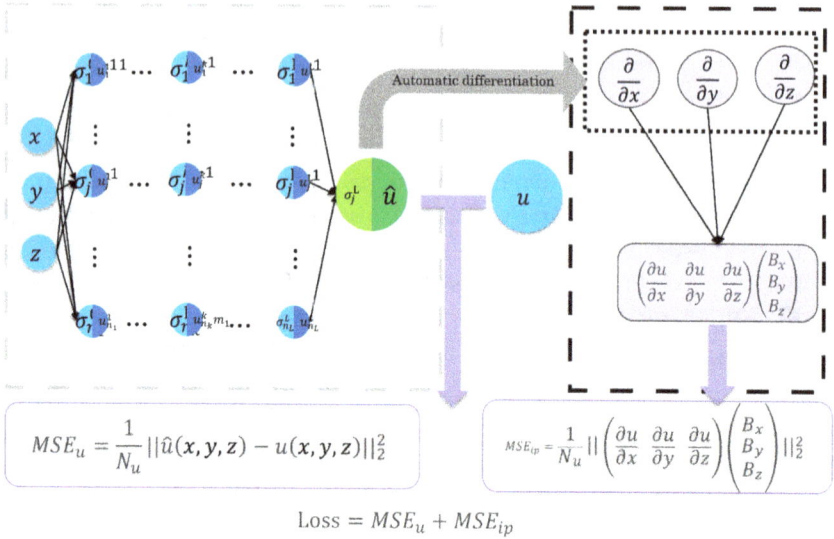

$$\text{Loss} = MSE_u + MSE_{ip}$$

Figure 3.8. Adding the inner product summation term to loss function.

as the interface of the plasma region [16]. The loss function obtained by adding the inner product term is as follows (Figure 3.8):

$$\text{Loss} = MSE_u + MSE_{ip}$$

A new error term composed of partial differential terms, MSE_{ip}, is added to the loss function. By adding objective functions (errors), so-called multi-objective optimization is converted into optimization with a single objective function. The strategy of adding multiple errors to the loss function in the NN as described earlier is the same as that for the PINNs proposed by the research group of M. Raissi *et al.* [28].

Figure 3.9 illustrates the visualized plasma region of the boundary surface, where a 3D spatial point in the fusion reactor is (x, y, z), a probability of a magnetic line that stays in the plasma region for a sufficiently long time passing through the 3D spatial point is $u(x, y, z)$, and the threshold is set to 0.5. The cases where the inner product summation term is added and not added are shown in the upper and lower columns, respectively. You can see that addition of the inner product summation term makes the region's surface smoother.

To visualize the plasma region in the fusion reactor, we train the NN with magnetic line group data extracted from the analysis results of the

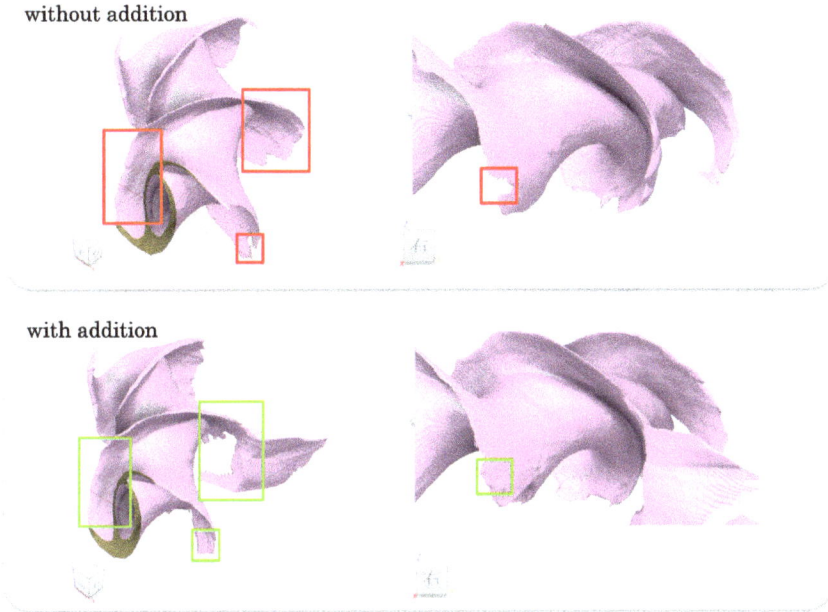

Figure 3.9. Difference in plasma region surface due to addition or no addition of inner product summation term.

electromagnetic field in the fusion reactor, and calculate the probability of the magnetic lines passing through the plasma region as volume data. In this chapter, we introduced how to do this. The NN model that represents the plasma region is characterized by adding a term consisting of the sum of the inner products between the gradient vector in the region and the vector in the magnetic line direction to the loss function. This method can be used to visualize the shape of a flow channel that envelops the streamlined data calculated from velocity fields, etc., and has wide-ranging applications. Chapter 4 deals with physical data and explains how to derive a PDE that describes the physical data using an NN.

3.6. Generation of a CAD model from point cloud data using point cloud data processing software

Point cloud data refer to data consisting only of coordinate points that are not associated with physical quantities. Point cloud data can be changed into physical data by labeling, as described in Section 2-7-5.

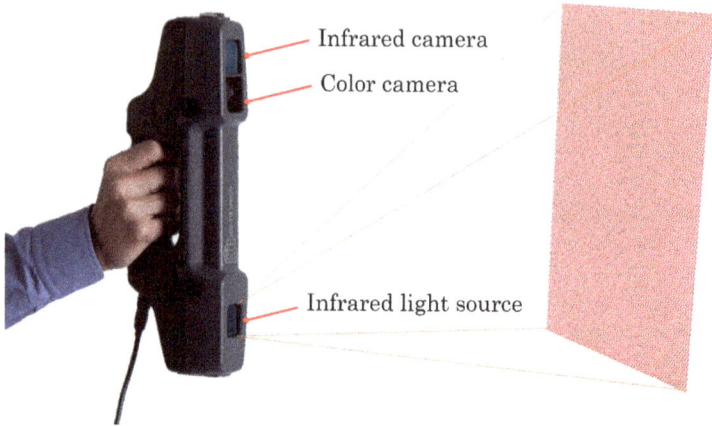

Figure 3.10. Generation of point cloud data using three-dimensional scanner (taken from https://ftr.co.jp/solution/3d-measurement/f6-smart/).

Recently, point cloud data have been generated in a variety of ways as follows:

- *Three-dimensional scanner*: A three-dimensional scanner is used to scan an object or environment to generate point cloud data (Figure 3.10). This includes LiDAR scanners, trigger scanners, and infrared scanners.
- *Cameras and lighting*: Cameras and lighting can be used to shoot an object or environment to generate point cloud data from the image. This includes point cloud extraction and multi-view stereopsis (MVS) (Figure 3.11).
- *Physical simulation*: Physical simulation can be used to generate point cloud data. This includes computational fluid dynamics (CFD) and finite element method (FEM) (Figure 3.12).
- *Physical data measurement by drones*: Drones can photograph large areas from the air and can be equipped with high-resolution cameras and sensors. This allows very detailed measurements of physical conditions, such as surface vegetation, land use, buildings, and structures. In addition, drones are equipped with GPS (Global Positioning System) and inertial measurement devices and can acquire data such as position information and flight speed. This allows you to get spatial data. In addition,

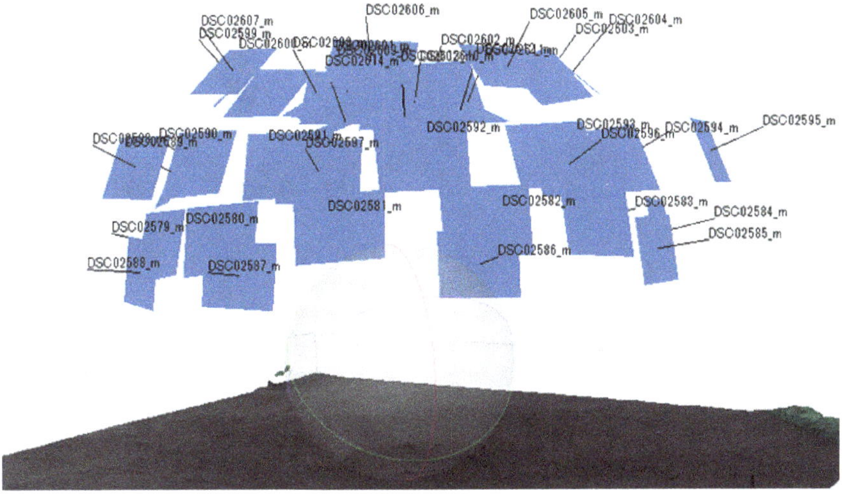

Figure 3.11. Point cloud data generation by camera (taken from https://aicam.jp/tech/sfmmvs).

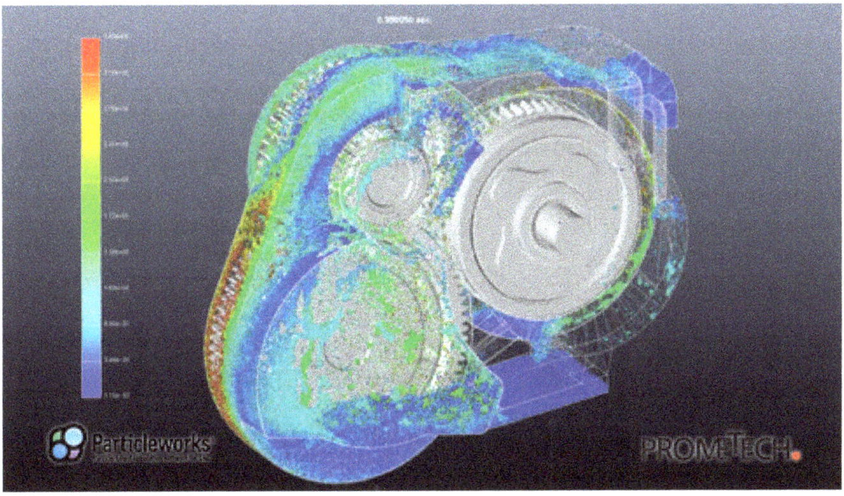

Figure 3.12. Point cloud data generation by physical simulation (taken from https://www.particleworks.com/case_study_ja.html).

Figure 3.13. Physical data measurement by drone (taken from https://viva-drone.com /drone-uav-surveying/).

drones can also be equipped with specialized sensors such as heat sensors and radiation meters. By using these sensors, physical data such as temperature distribution and radiation dose can be measured (Figure 3.13).
- *Programming*: To partially generate 3D data, programming can be used to generate point cloud data.

There are several approaches to generating a CAD model from point cloud data:

- *Poisson surface reconstruction (PSR)*: A method for generating triangular meshes from point cloud data to produce natural surface-like results (Figure 3.14).
- *Ball pivoting surface reconstruction (BPSR)*: A method for generating triangular meshes from point cloud data to generate a detailed surface (Figure 3.15).
- *Marching cubes (MC)*: A method for generating triangular meshes from point cloud data, which is usually used based on body element data, such as CT scan data (Figure 3.16).
- *Iterative closest point (ICP)*: A method for generating a CAD model from point cloud data, which is used to produce a shape similar to that of an existing CAD model (Figure 3.17).

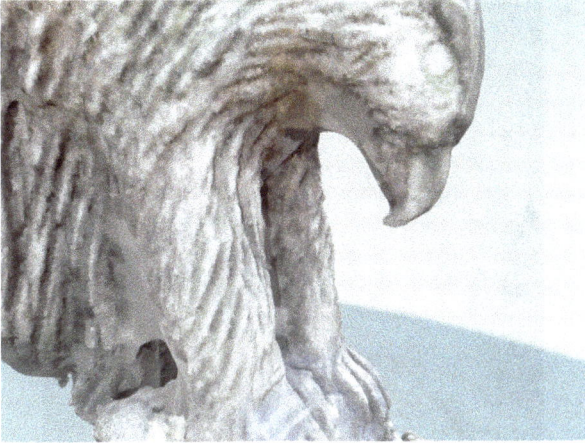

Figure 3.14. Triangular mesh generated from point cloud data by PSR (taken from https://medium.com/@sim30217/poisson-surface-reconstruction-7f319df6871c).

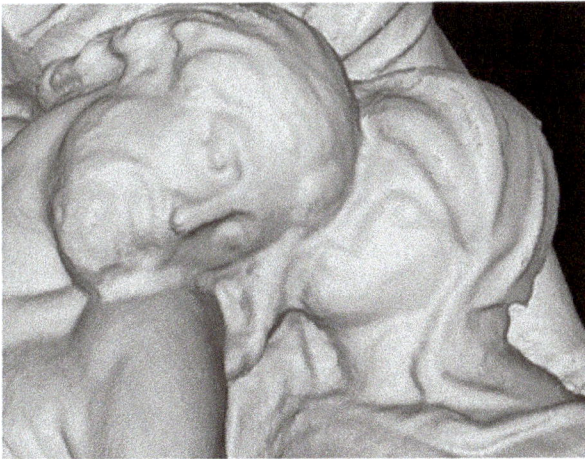

Figure 3.15. Triangle mesh generated from point cloud data by BPSR (taken from https://www.computer.org/csdl/journal/tg/1999/04/v0349/13rRUNvgz9y).

- *Multi-view stereopsis (MVS)*: A technique of three-dimensional reconstruction from point cloud data generated based on images captured by multiple cameras (Figure 3.18).

The selection of the above-mentioned approaches depends on the type of point cloud data and the required accuracy.

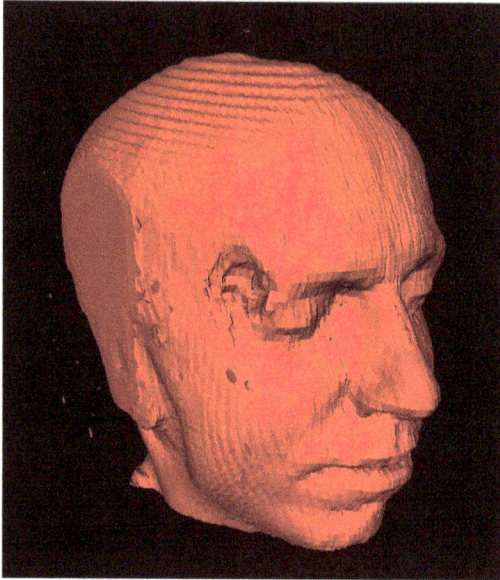

Figure 3.16. Triangle mesh generated from point cloud data by MC (taken from https://
en.wikipedia.org/wiki/Marching_cubes).

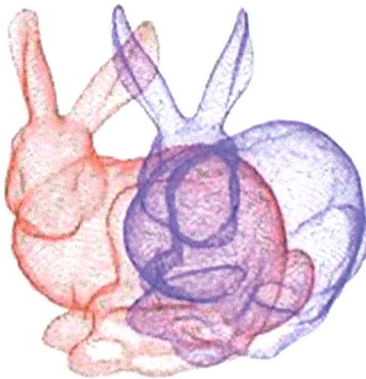

Figure 3.17. Triangle mesh generated from point cloud data by ICP (taken from http://
www.thothchildren.com/chapter/5c16831b41f88f26724b182f).

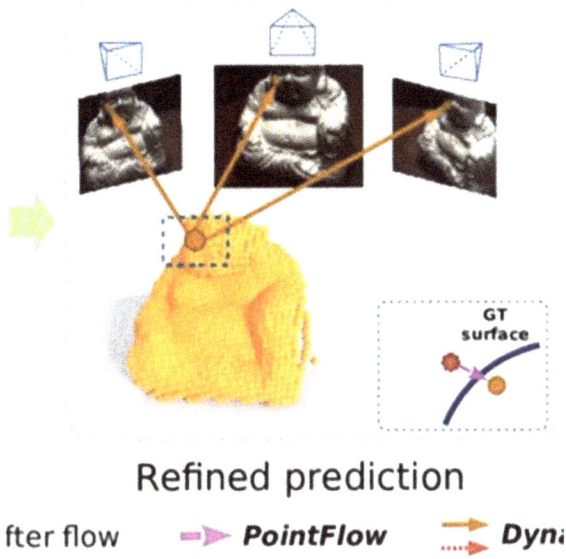

Figure 3.18. Point cloud data generated by MVS (taken from https://www.researchgate.net/figure/VA-Point-MVSNet-performs-multi-view-stereo-reconstruction-in-a-coarse-to-fine-fashion_fig1_340856447).

Chapter 4

Advanced Application Section

4.1. Derivation of PDEs describing physical data

Previous sections have thus far discussed the use of partial differentiation in the NN model. One of the applications discussed was the visualization of a curved surface which envelops a large number of magnetic field lines computed based on simulation results of electromagnetic fields in a fusion reactor. For effective visualization, a constraint is added to the loss function that the normal line of the curved surface is orthogonal to the magnetic flux density vector. This section describes how to derive PDEs that describe physical data tied to physical quantities using NN. As an application example, this section targets PDEs with exact solutions and describes how to find the original PDEs with high accuracy by regarding the results computed by randomly changing the space-time coordinates as pseudo-measurement data.

The PDE contains unknown multivariate functions and their partial derivatives. PDEs are used to formulate problems involving functions with several variables and are solved manually or using computer models. Incidentally, an ordinary differential equation (ODE) is expressed as a function of a single variable and its derivatives. This book assumes PDEs in which the time partial derivative of a physical quantity is represented by a linear regression model with a spatial partial derivative term. But this can be used for other cases as well.

Traditionally, PDEs were derived according to physical laws such as Fick's First Law or the continuity equation. For example, the advection–diffusion equation considers two types of material flows — advection and diffusion — and sums these flows into an integrated flow. The law of conservation is applied to this integrated flow to derive the advection–diffusion equation. This represents a deductive approach to PDE derivation. In contrast, this book describes the inductive approach.

Explanatory models that use existing PDEs are important to gain insights from the big data of interesting scientific events such as fluid, atmospheric, and cosmic phenomena, as well as various new occurrences such as COVID-19 infections.

Of course, it is also important to discover insights from everyday phenomena. For example, the use of computational fluid dynamics (CFD) is essential for the heat dissipation of laptop computers. Essentially, CFD is based on the convection–diffusion equation. For example, in thermo-fluid phenomena, if the heat dissipation effect is dominant in a laptop housing, the advection term can be omitted. This is a decision often adopted for dense board mounting. In this case, the derivation of the PDE from the thermo-fluid measurements plays an important role in determining whether an advection term is necessary. There have been significant technological advances in the simultaneous measurement of velocity, pressure, and temperature [18]. Thus, new PDEs that consist of essential partial differential terms may be derived from data measured with respect to the physical phenomena of interest.

PDE derivation techniques using cosmological big data are also expected to contribute to space science. In 1917, Einstein added a cosmological term, a type of partial differential term, to his equations to cancel out the effects of gravity, resulting in a static solution. In 1929, Edwin Hubble discovered the expansion of the universe, and Einstein withdrew the idea of the cosmological term. In 1998, observations of supernovae revealed that the expansion of the universe has continued to accelerate instead of slowing down over the past five billion years. In the near future, PDE derivation techniques should play an important role in providing evidence for the existence of a cosmological term in Einstein's equations.

Currently, research on PDE derivation from big data is in the initial technical research and development phase. This book assumes PDEs with exact solutions, and regards the exact solution data of the PDE at multiple discrete space-time points as pseudo-measurement data. A space-time model is needed to compute the temporal and spatial partial differential terms. For this reason, this book employs a space-time model based on NNs derived from physical data (Figure 3.5). NN represents the global approximation model. Traditionally, local interpolation models, such as finite difference and finite element models, were employed to evaluate partial differential terms in PDEs.

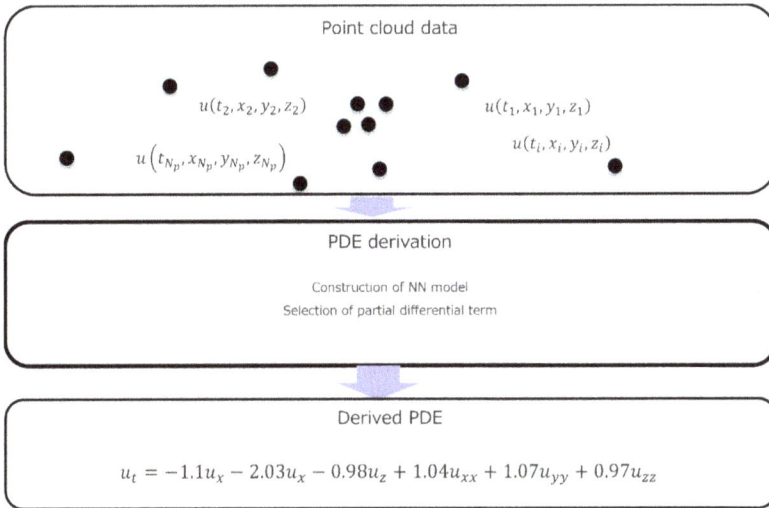

Figure 4.1. PDE derivation from physical data.

In PDE derivation, an NN model is constructed from the physical data previously described, and a partial differential term for a candidate component of the PDE is selected at discrete space-time points using automatic differentiation (Figure 4.1). Using partial differential terms, the coefficients are estimated by regularized regression [19]. Linear regression is a good predictive model for continuous variables, but it can be overtrained as the number of explanatory variables increases. Regularized regression is a linear regression that includes the concept of a regularization term to reduce overtraining. In regression analysis, a regularization term refers to a term introduced to take into account the complexity of the model, in addition to assessing the goodness of fit, to prevent overtraining. The regularization term is usually expressed in the form of an evaluation of the absolute value, squared value, and so forth, of variables representing weight parameters. Introducing a regularization term can give penalties to increased complexity of a model to decrease goodness of fit, thus preventing overtraining. Overtraining causes the absolute value of each standard partial regression coefficient to become very large. Here, the standard partial regression coefficient refers to the coefficient that indicates how much influence each explanatory variable has on the objective variable in a regression analysis. The standard partial regression coefficient is the coefficient after standardizing each explanatory variable (mean $= 0$, standard deviation $= 1$).

The standard partial regression coefficient is useful for comparing how important each variable is. It is often used because it makes evaluation easy. When the standard partial regression coefficient is larger, the fluctuation range of the objective variable comes finer and larger. To prevent this, a term representing the magnitude of the standard partial regression coefficient is included in the loss function. This is called a regularization term. This is expected to leave only the essential partial differential terms for the PDE.

A Python program that computes regularized regression employs a method that uses a machine learning library named scikit-learn. The following are sample codes for computing Lasso regularized regression using scikit-learn:

```
< Start >
from sklearn.linear_model import Lasso
import numpy as np
# Training data
X_train = np.array([[1, 2], [3, 4], [5, 6], [7, 8], [9, 10]])
y_train = np.array([1, 2, 3, 4, 5])
# Instantiate the model.
lasso = Lasso(alpha=0.1)
# Train the model
lasso.fit(X_train, y_train)
# Forecast
y_pred = lasso.predict(X_test)
< End >
```

Regularized regression can be easily computed using scikit-learn as shown here. The Lasso function in this sample program is a regression technique which uses the L1 norm as the regularization term. The alpha in this sample is a parameter that adjusts the strength of the regularization term. The smaller the alpha, the smaller the effect of the regularization term and the closer it is to normal regression. Conversely, as alpha increases, the impact of the regularization term intensifies, leading to more coefficients of variables approaching zero.

4.1.1. *Related research*

Complex nonlinear dynamics occur in many scientific and engineering disciplines. Therefore, clarifying the underlying differential equations directly from observations is a difficult task. The ability to symbolically model complex network systems is key to understanding such systems, and this remains an open question in many disciplines. It is essential to model the space-time continuum field from given physical data to evaluate partial derivatives and reveal the PDE. There are two approaches to representing the distribution of data in a field for accurate evaluation of partial differential terms: local interpolation and global approximation.

For the first approach (local interpolation), Bongard *et al.* proposed methods that can automatically generate symbolic equations for nonlinear coupled dynamical systems directly from time series data [27]. However, those methods were usable only to ODEs. Guo *et al.* studied the identification of a class of multi-scale space-time dynamical systems incorporating multiple spatial scales obtained from observations [28, 29]. This approach employed a local interpolation method that uses polynomials as basis functions, expands multilevel operators, and estimates spatial derivatives using the finite difference method (FDM). The coefficients of the polynomials may vary with respect to the spatial domain to represent multi-scale features related to system dynamics. They are approximated using B-spline wavelet multiple resolution analysis.

For the second approach (global approximation), Raissi *et al.* recently employed PINNs as the machine learning method. This means that the PDE is derived by introducing NN that learns to solve a supervised learning task while respecting the physical laws described by the general nonlinear PDE [30]. Raissi *et al.* introduced two main classes of algorithms to be used accordingly depending on whether the available data are scattered in space-time or placed in a fixed time snapshot: continuous-time model and discrete-time model. The effectiveness of these approaches has been demonstrated using a wide range of benchmark problems in mathematical physics, including conservation laws, incompressible fluid flow, and nonlinear shallow water wave propagation.

Hao Xu *et al.* proposed a data-driven method called DL-PDE, which combines NN and sparse regression methods to discover PDEs [31]. For physical problems, the NN is learned using available data, and sparse regression methods are used to identify sparse terms from the candidate library of partial differential terms that comprise the PDE.

Rudy *et al.* created a library of simple functions and partial derivatives that may appear in an unknown governing equation [32]. They then utilized the advantages of methods that promote sparsity and selected candidates that most accurately represent the data. Furthermore, Raissi and

Karniadakis [33] presented a framework for learning unknown parameters by introducing regularity between two consecutive time steps using a Gaussian process.

PDE is independent of the initial condition. Using the 3-D advection–diffusion equation, Koyamada *et al.* confirmed that the PDE derivation error decreases as the number of initial conditions increases [34].

Unfortunately, methods based on the second approach (global approximation) have not evaluated errors in the NN model which includes errors in the loss function and partial derivative terms, because it has not focused on exact solutions. This evaluation is important to find the optimal structure of the NN.

The PDE derivation described in this book is an inductive method that derives a PDE based solely on data collected at multiple space-time locations. In this method, two parameter optimization computations are performed. One is to build NN models from physical data, and the parameters to be optimized are the NN weights. That is, it searches for parameters that minimize the loss function. The other is to perform a regularized regression analysis using the partial differential terms computed for the NN model; the parameters to be optimized are the coefficients to the partial differential terms. That is, it searches for parameters (coefficient to the partial differential term) that minimize the regularized regression equation error.

4.1.2. *PDE derivation using regularized regression analysis*

NN model construction involves optimizing NN parameters through a computation process aimed at minimizing the loss function, which quantifies the disparity between the "measure" and the "approximation". The input is the space-time coordinates (x, y, z, t) and the output is the physical data u linked with that point, such as electric potential, temperature, and pressure (Figure 4.1). In this case, the loss function consists only of the mean squared error $E_u = \|\hat{u} - u\|_2^2 / N_p$ between the physical data and the data regressed by the NN model (hereinafter "the NN model error"). The circle labeled "σ" represents a neuron in each layer of the NN and its activation function is σ. Also, N_p represents the score. As for the NN, an FCNN where all neurons in each adjacent layer are connected is assumed. The NN model parameters add one additional count for the number of layers (L), the number of neurons per layer ($N_i, i = 1, \ldots, L$), the weights between neurons ($\boldsymbol{\omega} : w_{i,j}^k$, $i = 1, \ldots, N_k + 1, j = 1, \ldots, N_{k+1}, k = 1, \ldots, L$), and i. This is because the bias value for each neuron is considered a weight. In NN model building, the loss function is minimized by varying these multidimensional parameters.

The partial derivative terms are obtained from the NN model using the automatic differentiation described earlier. In this course, NN models are built using Google TensorFlow. TensorFlow is a platform for using machine learning methods such as NN and optimization techniques. Here, we used the GradientTape library in TensorFlow to compute the partial differential terms of the NN model.

Next, the essential partial differential terms that make up the PDE are selected. For the NN model, the regularized regression equation (RRE) is constructed using automatic differentiation (AD). If, as PDE, the first-order time partial derivative is expressed as a linear sum of spatial partial derivative terms, the relationship in given N_u space-time coordinates is expressed using the following regression equation:

$$\mathbf{u_t} = \boldsymbol{\Theta}\varepsilon$$

where $\mathbf{u_t}$ is a vector consisting of first-order time partial derivative values, $\boldsymbol{\Theta}$ is a matrix consisting of spatial partial derivative values at each level, and ε is the coefficient of their spatial partial derivative values. $\boldsymbol{\Theta}$ is the number of measured points, that is, the number of space-time coordinates N_P. There is no clear standard as to how far to include candidates for space-time partial differential terms in this regression equation. It is thus important to utilize visual analysis which can effectively use knowledge of experts. Next, constraints are added to this regression equation to construct RRE. One of the features of the RRE is that in the selection of the partial differential term, unnecessary partial differential terms are automatically eliminated, allowing the selection of essential partial differential terms. This is called the RRE error (E_{RRE}) because the value of RRE does not become zero during the construction of the NN model.

$$E_{\mathrm{RRE}} = \frac{1}{N_P}\|\boldsymbol{\Theta}\varepsilon - \mathbf{u_t}\|_2^2 + \lambda\frac{1}{N_\varepsilon}\|\varepsilon\|_2^2 \tag{4.1}$$

Using the coefficient $\hat{\varepsilon}$ of the spatial partial derivative obtained by minimizing E_{RRE}, the **PDE derivation error** can be defined by the following equation:

$$E_\varepsilon = \frac{1}{N_\varepsilon}\|\hat{\varepsilon} - \varepsilon\|_2^2 \tag{4.2}$$

where ε is known based on the assumption that the PDE is given. N_ε is the number of partial differential term candidates.

For example, as a PDE with an exact solution, if we focus on the following three-dimensional advection–diffusion equation, the PDE becomes as follows:

$$u_t = -\varepsilon_x u_x - \varepsilon_y u_x - \varepsilon_z u_z + \varepsilon_{xx} u_{xx} + \varepsilon_{yy} u_{yy} + \varepsilon_{zz} u_{zz} \qquad (4.3)$$

Subscripts indicate partial derivatives in time or space. Assuming that the candidate partial differential terms in the PDE are partial differential terms up to the second order, u_t is defined as follows:

$$u_t = \sum_{i=1}^{D} \varepsilon_{i,0} u_{x_i} + \sum_{i=1}^{D}\sum_{j=1}^{D} \varepsilon_{i,j} u_{x_i x_j} \qquad (4.4)$$

where u_{x_i} and u_t represent the partial differential terms due to x_i and time t, respectively, and each is given by automatic differentiation. D represents the number of dimensions of the spatial coordinate, which is three in this case. The PDE derivation is to estimate the value of the coefficient $\varepsilon_{i,j}$. Thus, regularized regression analysis is used to estimate these coefficients. This formulates the following least-squares problem for the RRE error:

$$\mathrm{argmin}_{\varepsilon} \left\{ \frac{1}{N_P} \sum_{N_P} \left(u_t - \sum_{i=1}^{D} \varepsilon_{i,0} u_{x_i} + \sum_{i=1}^{D}\sum_{j=i}^{D} \varepsilon_{i,j} u_{x_i x_j} \right)^2 + \lambda \frac{1}{N_\varepsilon} \|\varepsilon\|_2^2 \right\}$$

$$(4.5)$$

where ε denotes $\boldsymbol{\varepsilon} = [\varepsilon_{1,0} \cdots \varepsilon_{D,0}\, \varepsilon_{1,1} \cdots \varepsilon_{D,1}\, \varepsilon_{2,2} \cdots \varepsilon_{2,D} \cdots \varepsilon_{D,D}]^T$, and $\|\cdot\|_2^2$ denotes the vector sum of squares. The above problem is a convex optimization problem for which several practical algorithms have been proposed. Examples include the iterative shrinkage threshold algorithm, iteratively reweighted least squares (IRLS), and forward–backward splitting (FOBOS). Applying regularized regression can force certain coefficients to be set to zero.

4.1.3. *Definition of error*

It is important to define the error for the evaluation of PDE derivation methods. In addition to the NN model error (E_u), regularized regression equation error (E_{RRE}), and PDE derivation error (E_ε) described so far, the partial differential term error ($E_{u_{t.x..y..z...}}$) can be defined for PDEs with exact solutions as follows:

$$E_{u_{t.x..y..z...}} = \|\hat{u}_{t.x..y..z...} - u_{t.x..y..z...}\|_2^2 / N_P, \qquad (4.6)$$

where $\hat{u}_{t.x..y..z...}$ and $u_{t.x..y..z...}$ are partial differential terms computed from the NN model and exact solution, respectively. $\hat{u}_{t.x..y..z...} u_{t.x..y..z...}$ can be computed by automatic differentiation, and $u_{t.x..y..z...}$ can be computed analytically from the exact solution. For example, the exact solution of the advection–diffusion equation in three dimensions is as follows:

$$u(x,y,z,t) = \frac{1}{(4t+1)^{3/2}} e^{\left[-\frac{(x-\varepsilon_x t - c_x)^2}{\varepsilon_{xx}(4t+1)} - \frac{(y-\varepsilon_y t - c_y)^2}{\varepsilon_{yy}(4t+1)} - \frac{(z-\varepsilon_z t - c_z)^2}{\varepsilon_{zz}(4t+1)}\right]} \qquad (4.7)$$

where c_x, c_y, c_z is the constant determined by initial conditions. In the following, the partial differential term is denoted, as the case may be, as follows: $u_x = \frac{\partial u}{\partial x}$.

PDE derivation using NNs can reveal how each error relates to the NN's meta parameters, such as the number of layers and the number of neurons in each layer, and the discrete points in time and space (N_L, N_N, N_P). Visualization of the distribution of errors in the parameter space helps clarify the requirements for the NN structures that can maximize both NN model errors and regularized regression equation errors. As a result, errors in the NN model, PDE partial derivative term, regularized regression equation, and PDE derivation can be expressed as functions of the number of layers (N_L), that of neurons in the NN model (N_N), and that of discrete points (N_P).

In the case of the three-dimensional advection–diffusion equation derivation, the model is built using NN, followed by the selection of partial differential terms using regularized regression analysis. With respect to error evaluation, when the number of neurons per layer was kept the same and the number of discrete points was also kept constant, and each error was displayed in a heat map as the function of (N_L, N_N), there was a gap in points where each error was the smallest [25]. To solve this problem, the two parameter optimizations are integrated. That is, we developed a method to add E_u and E_{RRE} to establish a single objective function (loss function) and to search for an appropriate NN structure that minimizes it (Figure 4.2). In the PDE derivation, the PDE derivation error E_ε in Equation (4.2) also plays an important role, so if the correct PDE is known in advance, we will define the integrated error $E_{\text{int}} = E_u * E_{PDE} * E_\varepsilon$.

4.1.4. *Example of PDE derivation*

The PDE derivation methods described so far are applied to PDEs with exact solutions. The candidate partial differential terms are performed for two cases, the minimal and extended configurations, and the errors, E_{int}, are visualized for the NN parameters while changing the values. As NN

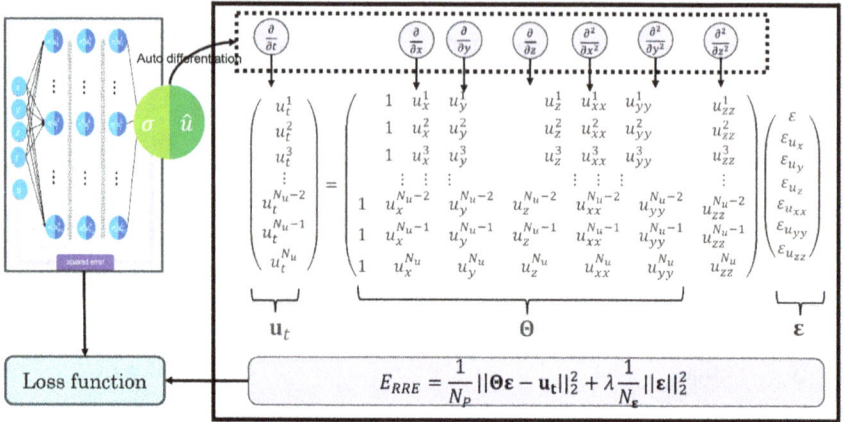

Figure 4.2. PDE derivation method using regularized regression.

parameters, we vary the number of neurons per layer and visualize them in parallel coordinates. The number of layers was fixed and the number of neurons in each layer was set as follows. In each layer, the number of neurons is classified into constant type, increase type, concave type, decrease type, and convex type. The four axes of the parallel coordinates represent the number of neurons (the width of the input layer matches the dimensionality of the equation, and that of the output layer is fixed at 1) and the last axis (error axis) represents the magnitude of the error E_{int}. We compute this for 10 PDE derivations and reveal the relationship between NN structure and error. In each NN structure type (for example, constant type), two structures are set according to the number of neurons. In this document, these 10 NN structures are hereafter referred to as N10. If two line segments intersect between the 4th axis and the error axis in the NN structure of the same classification, it can be interpreted that setting a large number of neurons reduces the error. The set consisting of the following 10 NNs is called N10:

# Constant type
layers1 = [4,40,40,40,40,1]
layers6 = [4,60,60,60,60,1]

# Increase type
layers2 = [4,10,20,30,40,1]
layers7 = [4,30,40,50,60,1]

# Concave type
layers3 = [4,40,30,30,40,1]
layers8 = [4,60,50,50,60,1]

# Decrease type
layers4 = [4,40,30,20,10,1]
layers9 = [4,60,50,40,30,1]

# Convex type
layers5 = [4,30,40,40,30,1]
layers10 = [4,50,60,60,50,1]

That is, N10 = {layers1, layers2, layers3, layers4, layers5, layers6, layers7, layers8, layers9, layers10}. In the following, we will present the calculation results in N10 and show the NN structure with the smallest error E_{int} for 8 PDEs, and then additionally visualize the relationship between the 10 errors and the NN structure.

4.1.4.1. *One-dimensional advection–diffusion equation*

The advection–diffusion (AD) equation is one of the equations used in physics and chemistry to mathematically model the flow of matter, energy, and other substances [35]. This equation describes the flow of matter and energy, taking into account both advection (flow) and diffusion (spread). Specifically, the advection equation describes the flow of matter, and the diffusion equation describes the diffusion of matter. This equation is used for many physical phenomena and engineering applications. In general, the basic equation can be expressed as follows:

$$\frac{\partial u}{\partial t} = \nabla \cdot (D\nabla u) - \nabla \cdot (\vec{v}u) + R \tag{4.8}$$

D is the diffusivity and v is the advection velocity. R is the generation term. The first series of experiments uses a spatial one dimensional AD equation. Assuming that the space can be represented in one dimension, the problem can be described as follows:

$$\text{PDE}: \frac{\partial u}{\partial t} + C\frac{\partial u}{\partial x} - D\frac{\partial^2 u}{\partial x^2} = 0 \tag{4.9}$$

Boundary condition: $u|_{x=0} = u_{BC}$; $u|_{x=\infty} = 0$

Initial condition: $u|_{t=0} = u_0(x)$

where $u_0(x)$ is the step function. Here, C is the x component of the advection vector. The domain of definition for this problem is $x, t \in [0, 10]$, and to evaluate the accuracy of the results, the exact solution with the boundary condition set to $u_{BC} = 1$ is used in the above equation.

$$u(x,t) = \frac{1}{2}\text{erfc}\left(\frac{x - Ct}{2\sqrt{Dt}}\right) + \frac{1}{2}\exp\left(\frac{Cx}{D}\right)\text{erfc}\left(\frac{x + Ct}{2\sqrt{Dt}}\right) \quad (4.10)$$

$$\text{erfc}(x) = \frac{2}{\sqrt{\pi}}\int_x^\infty \exp(-t^2)dt$$

where $\text{erfc}(x)$ is the complementary error function. In this experiment, $C = 1/2$, and $D = 1/14$ are used.

In the exact solution of the AD equation, the domain of definition of $x \in [0, 10]$ and $t \in [0, 10]$ is sampled, and $u(x,t)$ is computed based on the coordinate values, which is considered the pseudo-measurement data. For these data, the number of spatial observation points is $n_x = 256$, the number of temporal observation points is $n_t = 101$, and the dataset size is $n_d = 256856$. The results are shown in the top color map diagram in Figure 4.3. The partial differential term candidate library consists of four partial differential terms, including the correct partial differential term

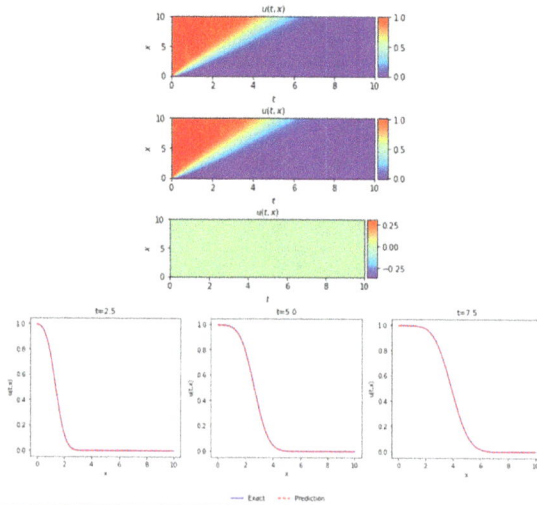

Correct PDE	$u_t + 1/2\, u_x - 1/14\, u_{xx}$ 0 $(u_t + 0.5u_x - 0.07142857u_{xx}\quad 0)$
Identified PDE (clean data)	$u_t - 0.534973u_x - 0.047054u_{xx} - 0.000043u_{xxxx}\quad 0$

Figure 4.3. Results of one-dimensional AD equation derivation with extended library.

(shown in bold), as follows:

$$\Phi = [\boldsymbol{u_{xx}}, \boldsymbol{u_x}, u_{xxx}, u_{xxxx}]$$

First, derivation of the AD equation is performed in N10. The activation function is the sigmoid function, sigmoid (x), and the number of datasets for training is randomly selected at $(n_d * 0.8)$ points. Results in which the NN has the minimum error are shown in the middle color map diagram in Figure 4.3. The lower color map diagram in Figure 4.3 shows the error distribution $(\hat{u}-u)$ between the NN model and the exact solution. The data distribution $\{u(x,t) \,|\, t = 2.5, 5.0, 7.5\}$ at fixed time is also shown below the error distribution. In addition, the bottom row of Figure 4.3 shows the derived PDEs, which indicate that the derivation error is not sufficient because of the large number of candidate partial derivative terms in the library.

The derivation error is reduced when the candidate partial differential terms in the library are limited to those contained in the original PDE. That is, if the partial differential term candidate library is configured as shown in $\Phi = [u_x, u_{xx}]$, Figure 4.4 shows the exact solution, NN model, error distribution, data distribution at fixed time, and derived PDE.

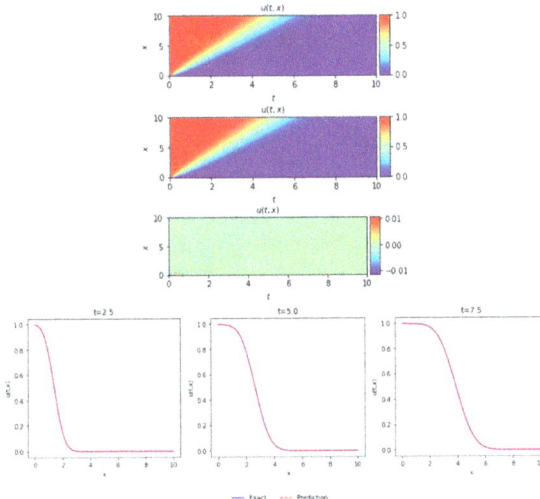

Correct PDE	$u_t + 1/2\,u_x - 1/14\,u_{xx}$ 0 $(u_t + 0.5u_x - 0.07142857u_{xx}$ 0 $)$
Identified PDE (clean data)	$u_t + 0.649396u_x - 0.001949u_{xx}$ 0

Figure 4.4. Results of one-dimensional AD equation derivation with minimum library.

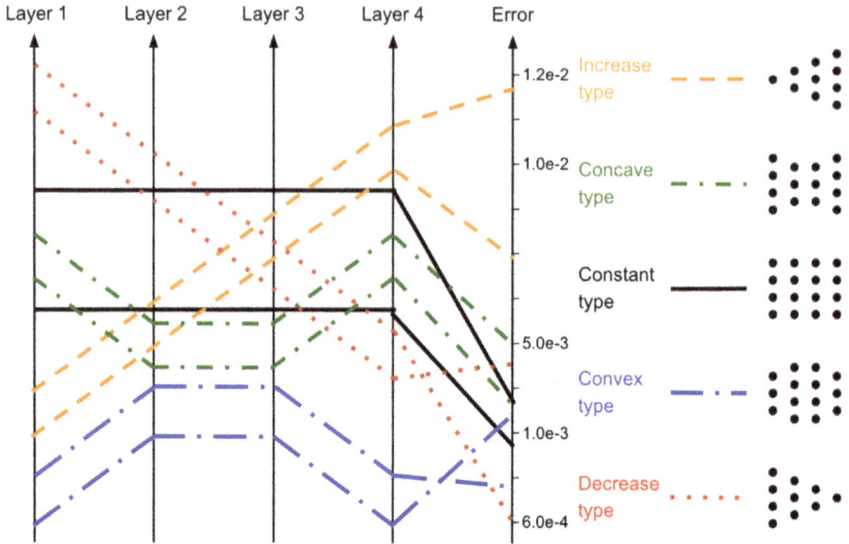

Figure 4.5. NN structure and E_{int} using extended library.

Figures 4.5 and 4.6 show the parallel coordinates of E_{int} when the approximate solution of the AD equation is computed in N10.

4.1.4.2. *One-dimensional KdV equation*

The one-dimensional KdV equation is a type of nonlinear partial differential equation that is expressed as follows:

$$u_t + \alpha u u_x + \beta u_{xxx} = 0 \qquad (4.11)$$

where $u(x,t)$ is the function of x and t, t is time, and x is space. Also, by applying the appropriate scale transformation to each variable and u, the coefficients can be recomputed as $\alpha = 6$, $\beta = 1$. Here, the partial derivative for each variable can then be expressed as lower right subscripts,

$$u_t + 6u u_x + u_{xxx} = 0. \qquad (4.12)$$

The KdV equation describes nonlinear wave phenomena such as shallow water waves [36]. This equation is a one-dimensional version of the Korteweg–De Vries equation and is used to describe nonlinear wave phenomena. Here, a nonlinear wave phenomenon refers to a phenomenon that occurs when nonlinearities that affect the medium through which the waves propagate appear. In contrast, a linear wave phenomenon refers to a phenomenon in which the effect of waves on a medium is linear.

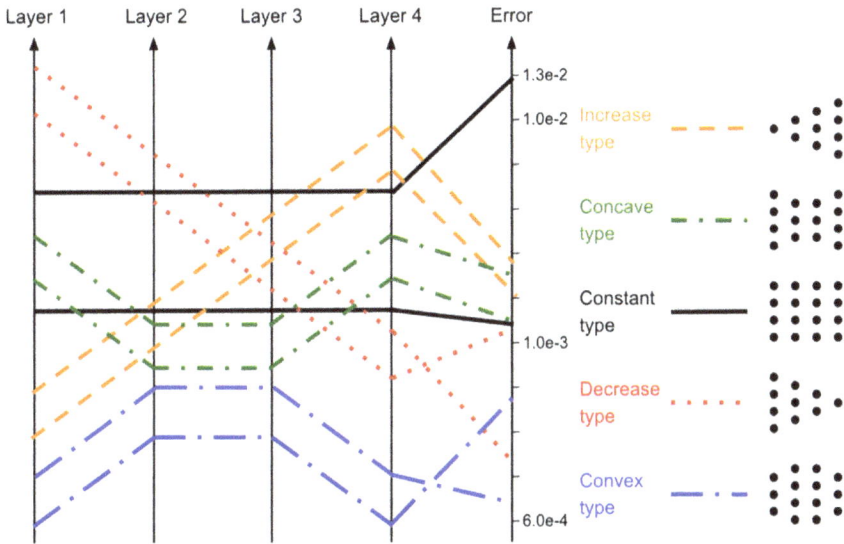

Figure 4.6. NN structure and E_{int} using minimum library.

This equation has local nonlinearity and nonlocal chromatic dispersion and is used to describe the nonlinear phenomenon of waves called soliton. It also plays an important role in quantum field theory, statistical mechanics, and mathematical physics. Here, chromatic dispersion refers to a physical phenomenon in which waves have different speeds depending on their wavelength (or frequency).

Specifically, depending on the nature of the medium in which the waves propagate, waves with shorter wavelengths may propagate faster than waves with longer wavelengths, or waves with longer wavelengths may propagate faster than waves with shorter wavelengths. This phenomenon is called chromatic dispersion. In an example of a wave traveling through water, the velocity of the wave depends on its wavelength. Waves with longer wavelengths slow down in shallow water, increasing the wave heights as the waves approach the shore. This is an example of chromatic dispersion in waves. Chromatic dispersion plays an important role in various physical phenomena such as optics, acoustics, and electromagnetic waves. Chromatic dispersion phenomena also play an important role in the analysis of nonlinear partial differential equations. Equations containing chromatic dispersion terms can describe wave deformation and wave interaction, among other phenomena.

Exact solutions exist for the one-dimensional KdV equation. Many exact solutions to this equation have been obtained through analytical methods. In particular, for initial value problems, Hirota's method and methods using the Miura transform are known and used to obtain exact solutions. There are also exact solutions for periodic and acyclic boundary conditions depending on their conditions. One of the best-known solutions of the KdV equation is called soliton. Soliton has an exact solution to the initial value problem, the waveform is invariant, and the wave speed and amplitude are constant regardless of time or space. A soliton solution has the following form:

$$u(x,t) = 2c_1 \left[\operatorname{sech}^2 \left(\frac{x - c_2 t - c_3}{c_4} \right) \right] \tag{4.13}$$

where c_1, c_2, c_3, and c_4 are arbitrary constants, c_1 is the soliton amplitude, c_2 is the soliton velocity, c_3 is the initial position of the soliton, and c_4 is the soliton width.

This solution is obtained by appropriately choosing c_1, c_2, c_3, or c_4 depending on the initial conditions in the initial value problem. The solution is also strongly stable against nonlinearity, the waveform is invariant, and the wave velocity and amplitude are constant regardless of time or space.

Soliton solutions play an important role in a variety of physical phenomena, including wave propagation, wave interaction, and the effects of nonlinearities.

In the exact solution of the KdV equation, the domain of definition of $x \in [0, 10]$ and $t \in [0, 10]$ is sampled, and $u(x,t)$ is computed based on their coordinates. The domain of the definition of x is divided into 256 parts, and a solution is computed for every $\Delta t = 0.1$ which is considered pseudo-measurement data. For these data, the number of spatial observation points is $n_x = 256$, the number of temporal observation points $n_t = 101$, and the dataset size $n_d = 256856$. The results are shown in the top color map diagram in Figure 4.7. The partial differential term candidate library consists of nine partial differential terms, including the correct partial differential term (shown in bold), as follows:

$$\Phi = [u_x, u_{xx}, \boldsymbol{u_{xxx}}, \boldsymbol{uu_x}, uu_{xx}, uu_{xxxx}, u^2 u_x, u^2 u_{xx}, u^2 u_{xxx}]$$

First, derivation of the KdV equation is performed in N10. The activation function is the sigmoid function, sigmoid (x), and the number of datasets for training is randomly selected at $(n_d * 0.8)$ points. Results in

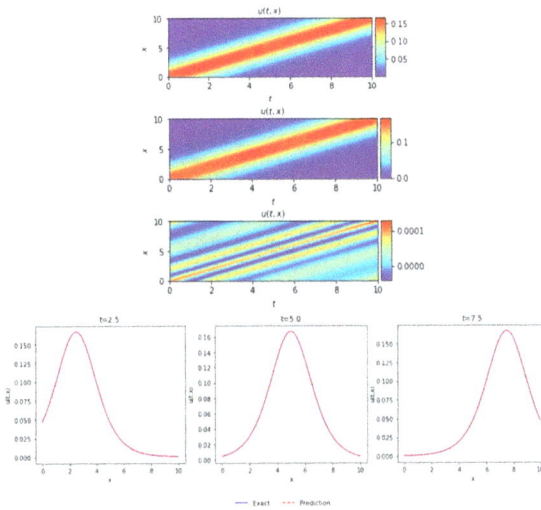

Correct PDE	$u_t + 6uu_x + u_{xxx} = 0$
Identified PDE (clean data)	$u_t + 1.000001u_x = 0$

Figure 4.7. Results of one-dimensional KdV equation derivation with extended library.

which the NN has the minimum error are shown in the middle color map diagram in Figure 4.7. The lower color map diagram in Figure 4.7 shows the error distribution $(\hat{u} - u)$ between the NN model and the exact solution. The data distribution $\{u(x,t) \mid t = 2.5, 5.0, 7.5\}$ at fixed time is also shown below the error distribution. In addition, the bottom row of Figure 4.7 shows the derived PDEs, which indicate that the derivation error is not sufficient because of the large number of candidate partial derivative terms in the library.

The derivation error is reduced when the candidate partial differential terms in the library are limited to those contained in the original PDE. That is, if the partial differential term candidate library is configured as shown in $\Phi = [uu_x, u_{xxx}]$, Figure 4.8 shows the exact solution, NN model, error distribution, data distribution at fixed time, and derived PDE.

Figures 4.9 and 4.10 show the parallel coordinates of E_{int} when the approximate solution of the KdV equation is computed in N10.

4.1.4.3. *One-dimensional Burgers' equation*

In physics, especially fluid dynamics, Burgers' equation is a second-order PDE describing one-dimensional nonlinear wave. This is the simplest

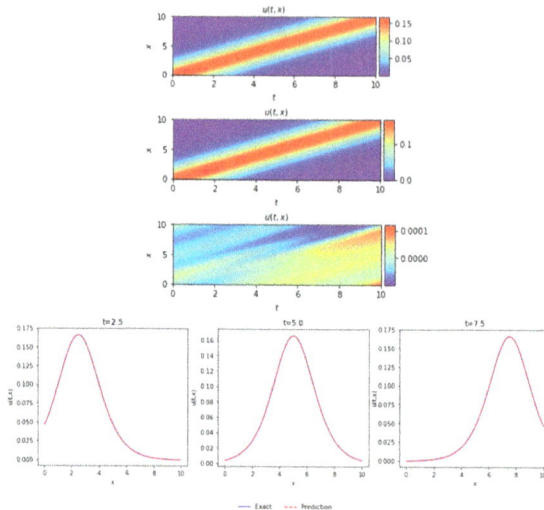

Correct PDE	$u_t + 6uu_x + u_{xxx} = 0$
Identified PDE (clean data)	$u_t + 18.019640uu_x + 1.004671u_{xxx} = 0$

Figure 4.8. Results of one-dimensional KdV equation derivation with minimum library.

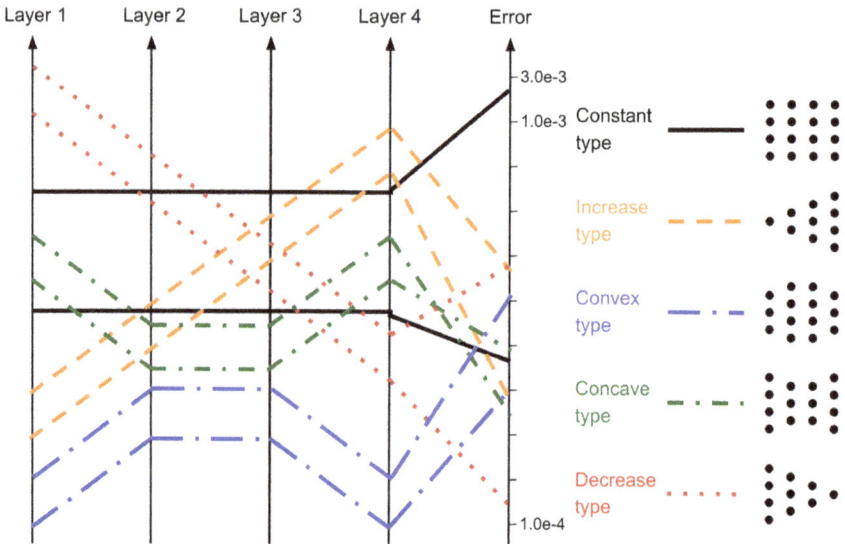

Figure 4.9. NN structure and E_{int} using extended library.

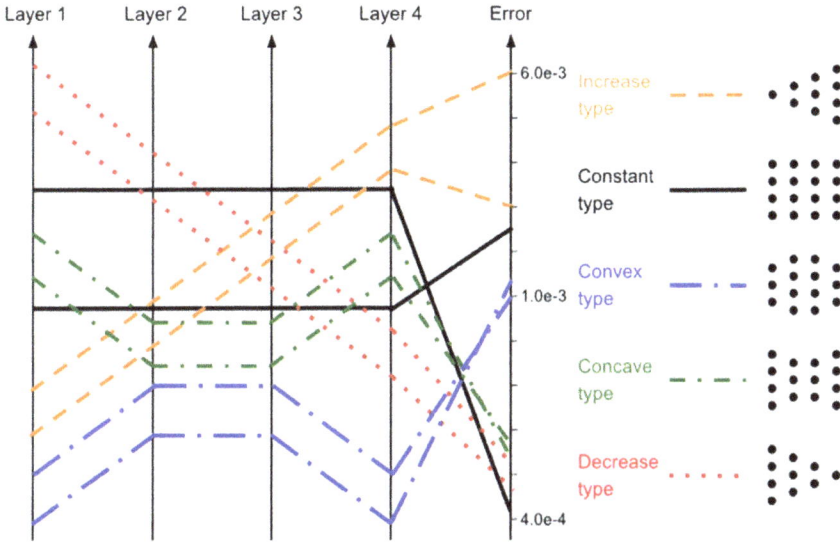

Figure 4.10. NN structure and E_{int} using minimum library.

nonlinear PDE that includes nonlinear and dissipative terms. It was first derived by Burgers and used to simulate a river channel [37]. Lighthill used it to simulate the shock wave phenomenon. He studied the existence of its global attractors, inertial manifolds, and approximate manifolds, and obtained many meaningful dynamical properties. It differs from the Navier–Stokes equations in that it removes the pressure gradient term and does not exhibit turbulent behavior. This occurs in various areas of engineering and applied mathematics, including fluid mechanics, nonlinear acoustics, gas dynamics, and traffic flows. Its one-dimensional form is given as follows:

$$u_t + uu_x + \alpha u_{xx} = 0 \tag{4.14}$$

where α is the diffusion coefficient, which, in this example, is set to 0.1. The initial condition is $u(x,0) = -\frac{2a}{\phi}\frac{\partial\phi}{\partial x} + 4$. Here, it becomes $\phi = e^{\left(\frac{-x^2}{4a}\right)} + e^{\left(\frac{-(x-2\pi)^2}{4a}\right)}$. The exact solution in this case is as follows:

$$u(x,t) = -\frac{0.2}{\phi}\frac{\partial\phi}{\partial x} + 4 \tag{4.15}$$

$$\phi = e^{\left(\frac{-(x-4t)^2}{0.4(t+1)}\right)} + e^{\left(\frac{-(x-4t-2\pi)^2}{0.4(t+1)}\right)}$$

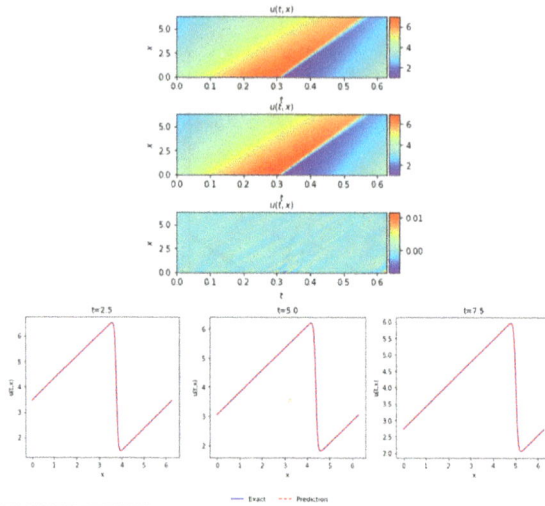

Figure 4.11. Results of one-dimensional Burgers' derivation with extended library.

Correct PDE	$u_t + uu_x + 0.1u_{xx} = 0$
Identified PDE (clean data)	$u_t - 0.004922u - 0.101164u_{xx} - 0.000198u_{xx} + 1.009398uu_x + 0.000454uu_{xx} + 0.000096uu_{xx}$ $- 0.002101uuu_x - 0.000011uuu_{xxx} = 0$

Unlike the previous example, the dataset for training the NN model is obtained by simulation using traditional spectral methods. The above Burgers' equation is integrated from the start time $t = 0$ to the end time $t = \frac{\pi}{5}$. This is accomplished by Fourier spectral discretizations of 101 modes in the spatial domain $x \in [0, 2\pi]$ and an explicit Runge–Kutta time integrator of the first order with time step size $\pi/500$. A solution for each $\Delta t = \frac{\pi}{500}$ is recorded to obtain 100 observation points. These are $n_x = 101$, $n_t = 100$, and $n_d = 10100$ for these data.

The results are shown in the top color map diagram in Figure 4.11. The partial differential term candidate library is constructed as follows:

$$\Phi = [u_x, \boldsymbol{u_{xx}}, u_{xxx}, \boldsymbol{uu_x}, uu_{xx}, uu_{xxxx}, u^2 u_x, u^2 u_{xx}, u^2 u_{xxx}]$$

First, derivation of Burgers' equation is performed in N10. Results in which the NN has the minimum error are shown in the middle color map diagram in Figure 4.11. The lower color map diagram in Figure 4.11 shows the error distribution $(\hat{u} - u)$ between the NN model and the exact solution.

The data distribution $\{u(x, t) \mid t = 2.5, 5.0, 7.5\}$ at fixed time is also shown below the error distribution. In addition, the bottom row of

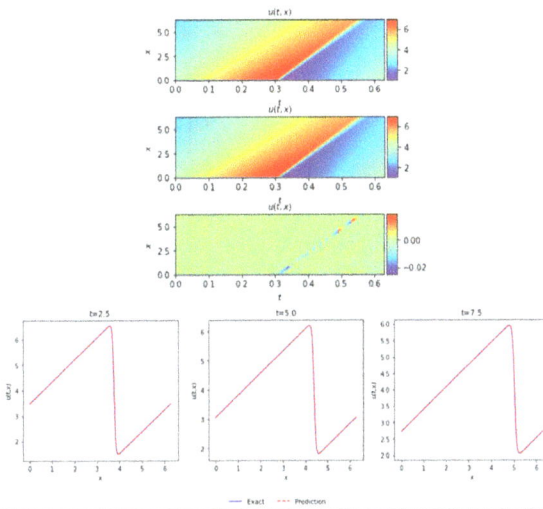

Correct PDE	$u_t + uu_x + 0.1u_{xx} = 0$
Identified PDE (clean data)	$u_t + 0.999757uu_x - 0.099821u_{xx} = 0$

Figure 4.12. Results of one-dimensional Burgers' derivation with minimum library.

Figure 4.11 shows the derived PDE, but the derivation error is not sufficient because of the large number of candidate partial derivative terms in the library.

The derivation error improves when the candidate partial differential terms in the library are minimized to only the exact ones. That is, if the partial differential term candidate library is configured as shown in $\Phi = [uu_x, u_{xx}]$, Figure 4.12 shows the exact solution, NN model, error distribution, data distribution at fixed time, and derived PDE.

Figures 4.13 and 4.14 show the parallel coordinates of E_{int} when the approximate solution of Burgers' equation is computed in N10.

4.1.4.4. *One-dimensional Poisson's equation*

Poisson's equation is a type of partial differential equation used to describe the distribution of certain physical quantities in space-time. Specifically, Poisson's equation is expressed as follows:

$$\nabla^2 u = -\rho \tag{4.16}$$

where u is a scalar field representing the distribution of physical quantities and ρ is a source term such as charge density or mass density in space.

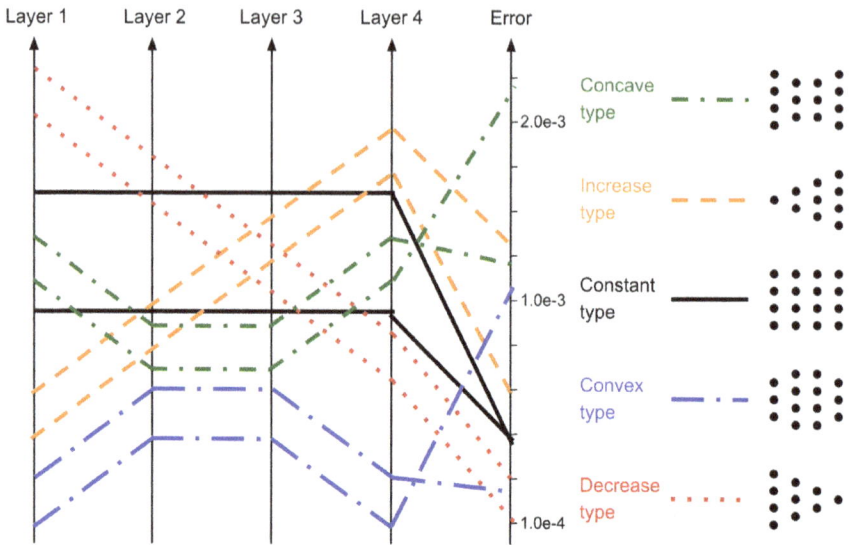

Figure 4.13. NN structure and E_{int} using extended library.

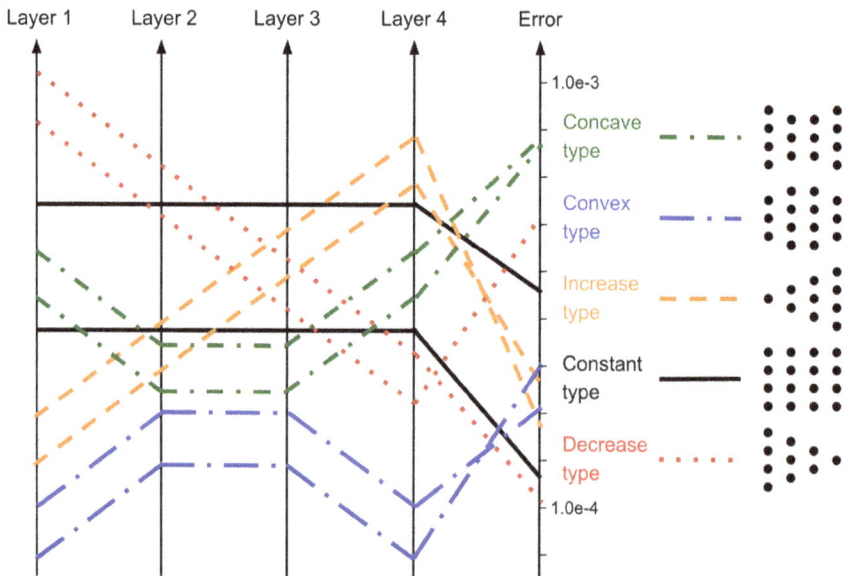

Figure 4.14. NN structure and E_{int} using minimum library.

∇^2 is the Laplacian operator and is defined for the three spatial coordinates of x, y, z by the following equation:

$$\nabla^2 = \partial^2/\partial x^2 + \partial^2/\partial y^2 + \partial^2/\partial z^2$$

Poisson's equation is widely used to describe potential distributions such as electric and gravitational fields [38]. Specifically, when an electric charge distribution is presented, Poisson's equation is used to find the potential distribution. Poisson's equation also plays an important role in areas such as fluid mechanics and heat conduction. In one-dimensional space, the one-dimensional Poisson's equation can be written as

$$\frac{\partial^2 u}{\partial x^2} - t = 0 \qquad (4.17)$$

When the definition domain and boundary conditions are set as follows,

$$(x, t) \in [0, 1] \times [0, 1]$$

$$u|_{x=-1} = 0; \ u|_{x=1} = 0$$

the exact solution is obtained as follows:

$$u = \frac{x^2}{2}t - \frac{t}{2} \qquad (4.18)$$

In the exact solution of Poisson's equation, the domain of definition of $x \in [0, 1]$ and $t \in [0, 1]$ is sampled, and $u(x, t)$ is computed based on their coordinates, which are considered pseudo-measurement data. The partial differential term candidate library is constructed as follows:

$$\Phi = [u_x, \mathbf{u_{xx}}, u_{xxx}]$$

The results are shown in the top color map diagram in Figure 4.15.

First, derivation of Poisson's equation is performed in N10. Results in which the NN has the minimum error are shown in the middle color map diagram in Figure 4.15. The lower color map diagram in Figure 4.15 shows the error distribution $(\hat{u} - u)$ between the NN model and the exact solution.

The data distribution $\{u(x, t) \mid t = 0.25, 0.5, 0.75\}$ at fixed time is also shown below the error distribution. In addition, the bottom row of Figure 4.15 shows the derived PDE, but the derivation error is not sufficient because of the large number of candidate partial derivative terms in the library.

The derivation error improves when the candidate partial differential terms in the library are minimized to only the exact ones. That is, if the partial differential term candidate library is configured as shown in

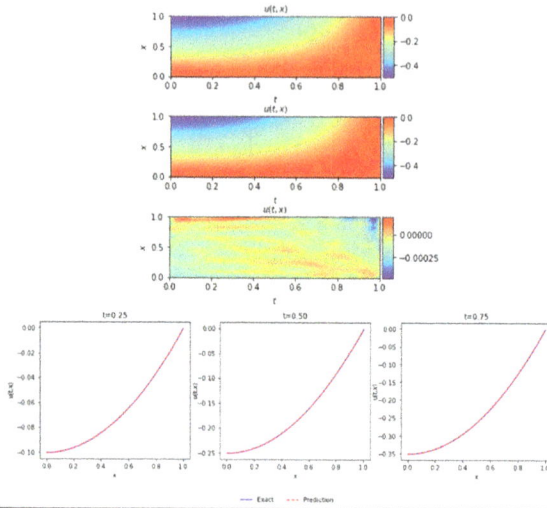

Correct PDE	$u_{xx} - t = 0$
Identified PDE (clean data)	$0.068095u + 0.947966u_x - 0.020915u_{xx} - t = 0$

Figure 4.15. Results of one-dimensional Poisson's equation derivation with extended library.

$\Phi = [u_{xx}]$, Figure 4.16 shows the exact solution, NN model, error distribution, data distribution at fixed time, and derived PDE.

Figures 4.17 and 4.18 show the parallel coordinates of E_{int} when the approximate solution of Poisson's equation is computed in N10.

4.1.4.5. *One-dimensional heat conduction equation*

The heat conduction equation is a partial differential equation that describes how heat is transferred inside a material. The heat conduction equation is used to describe the temperature distribution of a material.

In one dimension, the heat conduction equation is expressed as follows:

$$\frac{\partial u}{\partial t} - D\frac{\partial^2 u}{\partial x^2} = 0 \tag{4.19}$$

where u represents the temperature distribution inside a material, t represents time, and x the coordinates in one-dimensional space. D is the thermal diffusivity, which depends on the thermal conduction properties of the material.

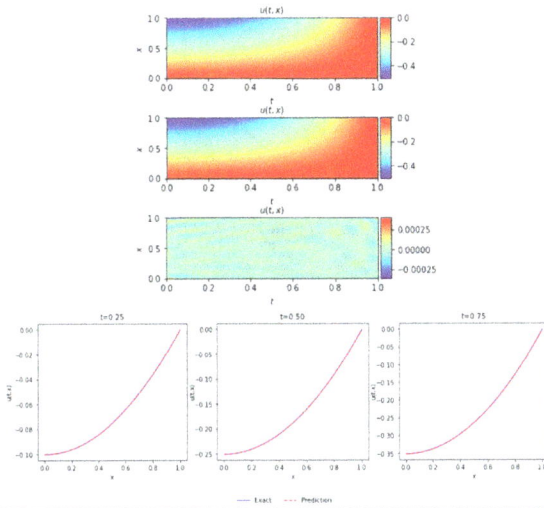

Figure 4.16. Results of one-dimensional Poisson's equation derivation with minimum library.

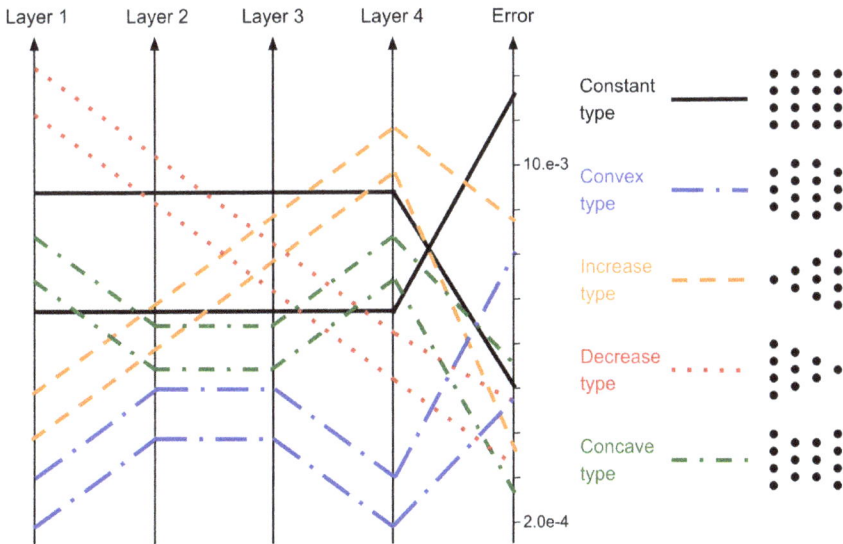

Figure 4.17. NN structure and E_{int} using extended library.

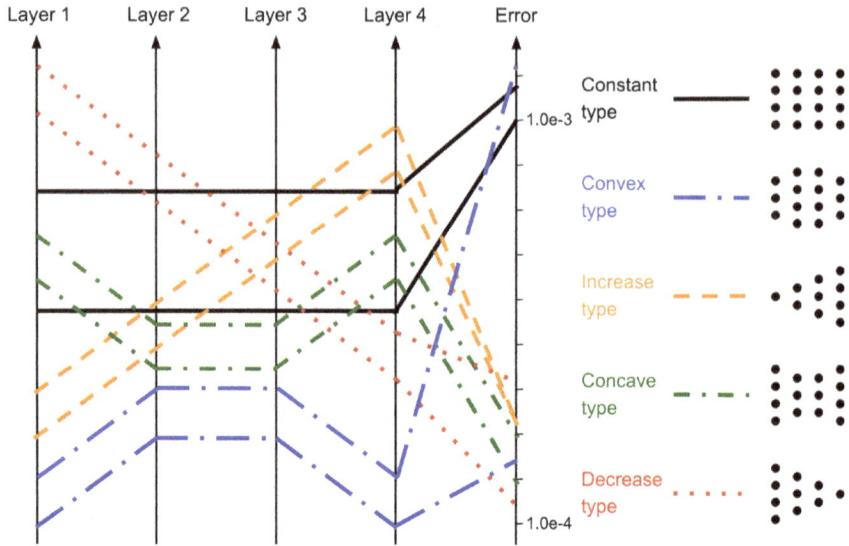

Figure 4.18. NN structure and E_{int} using minimum library.

The heat conduction equation describes heat conduction phenomena inside a material and is used to analyze thermal diffusion phenomena and the propagation of temperature changes [39]. This equation plays an important role in industrial applications such as heat treatment and processing of materials, because it describes the time evolution of temperature changes inside a material. Heat conduction equations are also sometimes used in other fields such as geophysics and astronomy.

When the definition domain and boundary conditions are set as follows,

$$(x, t) \in [0, 10] \times [0, 10]$$

$$u|_{t=0} = 1; \ u|_{x=1} = 0$$

the exact solution is obtained as follows:

$$u = \text{erfc}\left(\frac{x}{2\sqrt{Dt}}\right) \tag{4.20}$$

In the exact solution of the heat conduction equation, the domain of definition of $x \in [0, 10]$ and $t \in [0, 10]$ is sampled, and $u(x, t)$ is computed based on their coordinates, which are considered pseudo-measurement data.

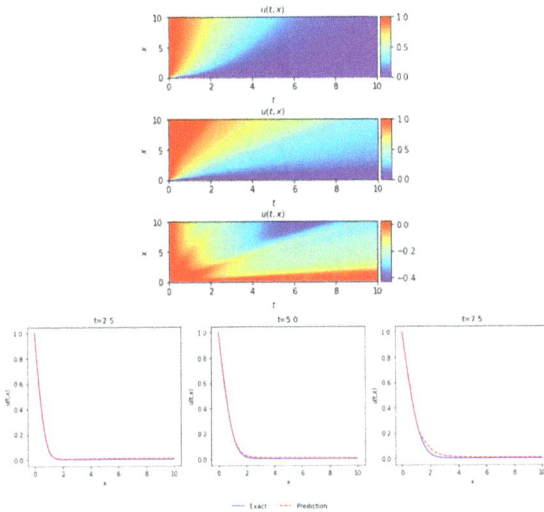

Figure 4.19. Results of one-dimensional heat conduction equation derivation with extended library.

The partial differential term candidate library is constructed as follows:

$$\Phi = [u_x, \boldsymbol{u_{xx}}, u_{xxx}]$$

The results are shown in the top color map diagram in Figure 4.23.

First, derivation of the heat conduction equation is performed in N10. Results in which the NN has the minimum error are shown in the middle color map diagram in Figure 4.19. The lower color map diagram in Figure 4.19 shows the error distribution $(\hat{u} - u)$ between the NN model and the exact solution.

The data distribution $\{u(x, t) \mid t = 2.5, 5, 7.5\}$ at fixed time is also shown below the error distribution. In addition, the bottom row of Figure 4.19 shows the derived PDE, but the derivation error is not sufficient because of the large number of candidate partial derivative terms in the library.

The derivation error improves when the candidate partial differential terms in the library are minimized to only the exact ones. That is, if the partial differential term candidate library is configured as shown in $\Phi = [u_{xx}]$, Figure 4.20 shows the exact solution, NN model, error distribution, data distribution at fixed time, and derived PDE.

Figures 4.21 and 4.22 show the parallel coordinates of E_{int} when the approximate solution of the heat conduction equation is computed in N10.

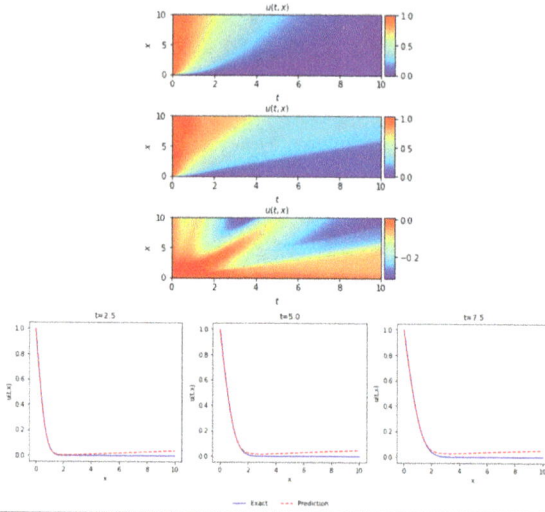

Correct PDE	$u_t - u_{xx} = 0$
Identified PDE (clean data)	$u_t + 0.002241u_{xx} = 0$

Figure 4.20. Results of one-dimensional heat conduction equation derivation with minimum library.

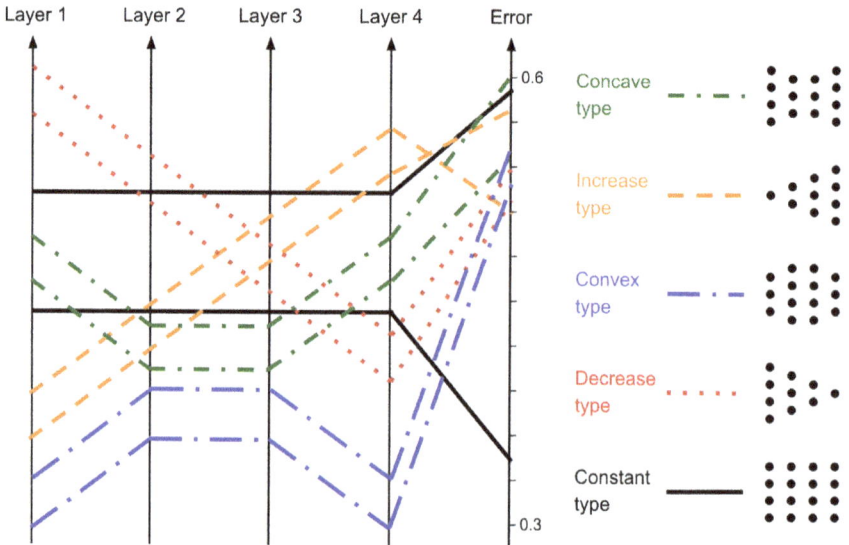

Figure 4.21. NN structure and E_{int} using extended library.

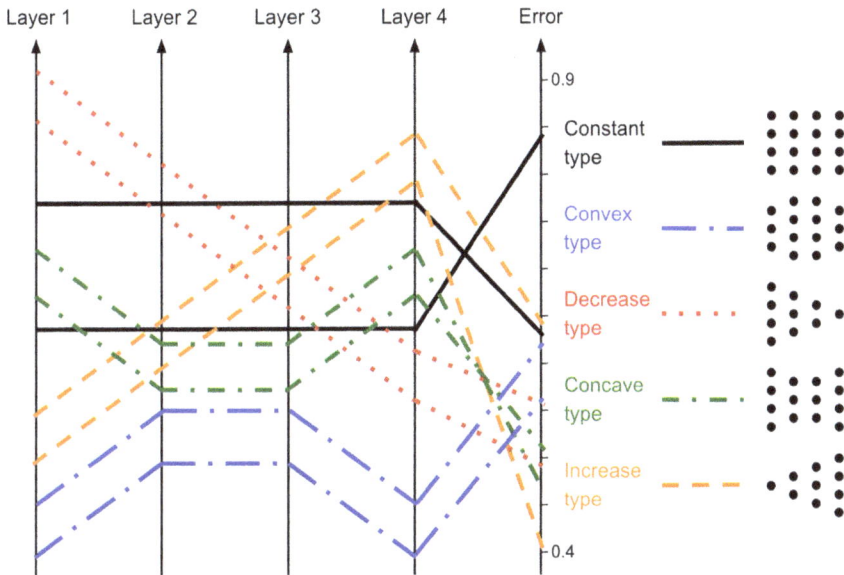

Figure 4.22. NN structure and E_{int} using minimum library.

4.1.4.6. *One-dimensional wave equation*

The wave equation is a type of partial differential equation for describing wave phenomena and can be expressed as

$$\frac{\partial^2 u}{\partial t^2} - c^2 \frac{\partial^2 u}{\partial x^2} = 0 \tag{4.21}$$

where u represents wave amplitude, t represents time, and x represents spatial coordinates. c is the wave velocity, which depends on the properties of the medium through which the wave propagates.

The wave equation is used to describe various types of wave phenomena, such as sound waves, electromagnetic waves, water waves, and seismic waves. This equation describes the temporal evolution of waves and thus helps analyze phenomena such as wave propagation, reflection, and interference [40].

The wave equation is also used in a wide range of applications in physics. Examples include quantum mechanical analysis of electromagnetic waves, acoustics, hydraulics, meteorology, seismology, and electrical circuit theory. The wave equation has been applied in a variety of fields and is the

Analysis and Visualization of Discrete Data

foundation of modern physics and engineering. In one-dimensional problems, the wave equation can also be understood as an oscillation.

When the definition domain and boundary conditions are set as follows,

$$(x, t) \in [0, 1] \times [0, 1]$$

$$u|_{t=0} = \sin(3\pi x); u|_{x=0} = 0$$

the exact solution is obtained as follows:

$$u = \cos(3\pi ct) \sin(3\pi x) \tag{4.22}$$

In the exact solution of the heat wave equation, the domain of definition of $x \in [0, 1]$ and $t \in [0, 1]$ is sampled, and $u(x, t)$ is computed based on their coordinates, which are considered pseudo-measurement data. The partial differential term candidate library is constructed as follows:

$$\Phi = [u_x, \boldsymbol{u_{xx}}, u_{xxx}]$$

The results are shown in the top color map diagram in Figure 4.23. First, derivation of the heat conduction equation is performed in N10. Results in which the NN has the minimum error are shown in the middle color map diagram in Figure 4.23. The lower color map diagram in

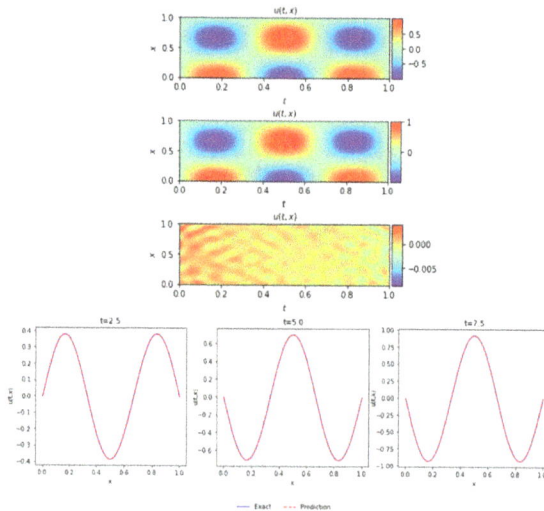

| Correct PDE | $u_{tt} - 0.25u_{xx} \quad 0$ |
| Identified PDE (clean data) | $u_{tt} + 2.021115u_x - 0.000768u_{xx} + 0.010821u_{xxx} \quad 0$ |

Figure 4.23. Results of one-dimensional wave equation derivation with extended library.

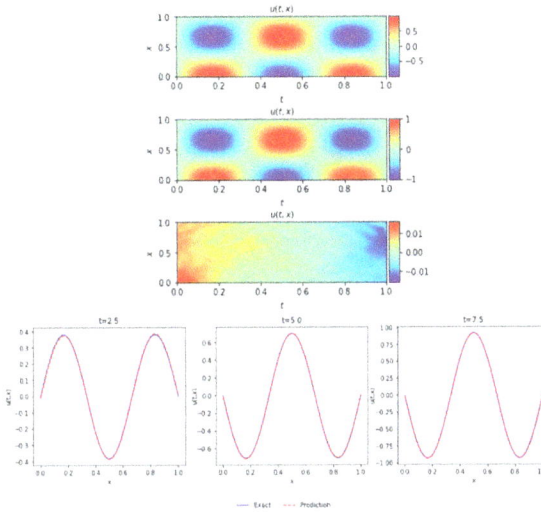

Correct PDE	$u_{tt} - 0.25u_{xx} = 0$
Identified PDE (clean data)	$u_{tt} - 0.001593u_{xx}$ 0

Figure 4.24. Results of one-dimensional wave equation derivation with minimum library.

Figure 4.23 shows the error distribution $(\hat{u} - u)$ between the NN model and the exact solution.

The data distribution $\{u(x,t)\,|\,t = 0.25, 0.5, 0.75\}$ at fixed time is also shown below the error distribution. In addition, the bottom row of Figure 4.23 shows the derived PDE, but the derivation error is not sufficient because of the large number of candidate partial derivative terms in the library.

The derivation error improves when the candidate partial differential terms in the library are minimized to only the exact ones. That is, if the partial differential term candidate library is configured as shown in $\Phi = [u_{xx}]$, Figure 4.24 shows the exact solution, NN model, error distribution, data distribution at fixed time, and derived PDE.

Figures 4.25 and 4.26 show the parallel coordinates of E_{int} when the approximate solution of the wave equation is computed in N10.

4.1.4.7. *Two-dimensional linear stress analysis*

In the PDEs discussed so far, the target physical phenomena were limited to those related to fluids. So, this section will cover the discussion of stress analysis in solids. Unlike fluid analysis, stress analysis often uses the

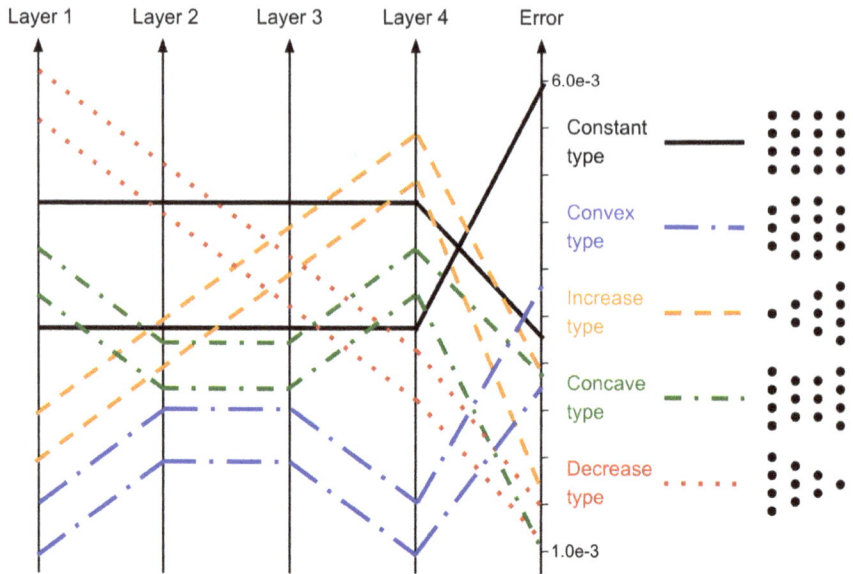

Figure 4.25. NN structure and E_{int} using extended library.

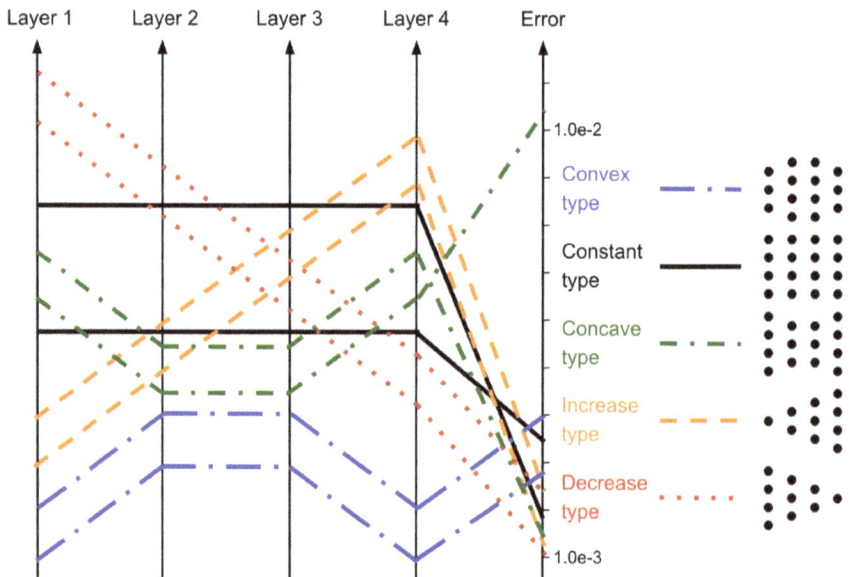

Figure 4.26. NN structure and E_{int} using minimum library.

displacement method and seems less likely to explicitly face PDEs. This
section describes PDEs on stress analysis assuming a plane stress state.

Stress analysis is used to understand the elastic response of structures.
Elasticity refers to the ability of a structure to return to its original state
after being subjected to load or deformation. The linear stress analysis
problem described here applies to cases where loads or strains applied to a
structure increase gradually. It also assumes that the stresses and strains
applied to the structure increase linearly. In contrast, it can be shown
that as the stress and strain applied to a structure increase, the response
also increases. The linear stress analysis problem can be used to determine the stress concentration in a structure which corresponds the loads
applied to the structure. This allows analyses of structural vibration and
fracture.

When solving linear stress analysis problems, the virtual work princi-
ple is often used in FEM [41]. FEM is a method for numerically analyzing
physical systems on a computer. The physical system to be analyzed is
divided into a finite number of elements, and the equilibrium state of the
entire system is obtained by solving the local equilibrium equations for
each element. In FEM, the virtual work method is used to find the equilibrium equations for each element. The equilibrium equation describes the
equilibrium between the resultant and reaction forces in an element. The
virtual work method is to find the forces applied to the element, compute
the effect of those forces on the element, and check to see if the results of
those computations are zero.

In FEM, the virtual work method is used to solve local equilibrium
equations. By combining those local equilibrium equations, the equilibrium
state of an entire system can be obtained. For this reason, the linear stress
analysis problems have not addressed PDEs.

A linear stress analysis problem can be expressed as a PDE considering
that it is described by three equations as follows:

Equilibrium equation:

$$
\underbrace{\left\{ \begin{array}{ccc} \dfrac{\partial}{\partial x} & 0 & \dfrac{\partial}{\partial y} \\[2mm] 0 & \dfrac{\partial}{\partial y} & \dfrac{\partial}{\partial x} \end{array} \right\}}_{\nabla*T} \underbrace{\left\{ \begin{array}{c} \sigma_x \\ \sigma_y \\ \sigma_{xy} \end{array} \right\}}_{\vec{\sigma}} + \underbrace{\left\{ \begin{array}{c} f_x \\ f_y \end{array} \right\}}_{\vec{f}} = \left\{ \begin{array}{c} 0 \\ 0 \end{array} \right\} \tag{4.23}
$$

Relationship between strain and displacement:

$$
\left\{
\begin{array}{c}
\varepsilon_x \\
\varepsilon_y \\
\varepsilon_{xy}
\end{array}
\right\}
=
\left\{
\begin{array}{c}
\dfrac{\partial u}{\partial x} \\[2mm]
\dfrac{\partial v}{\partial y} \\[2mm]
\dfrac{\partial u}{\partial y} + \dfrac{\partial v}{\partial x}
\end{array}
\right\}
=
\left\{
\begin{array}{cc}
\dfrac{\partial}{\partial x} & 0 \\[2mm]
0 & \dfrac{\partial}{\partial y} \\[2mm]
\dfrac{\partial}{\partial y} & \dfrac{\partial}{\partial x}
\end{array}
\right\}
\left\{
\begin{array}{c}
u \\
v
\end{array}
\right\}
\tag{4.24}
$$

$\underbrace{}_{\vec{\varepsilon}}$ $\underbrace{}_{\nabla^*}$ $\underbrace{}_{\vec{u}}$

Constitutive equation:

$$
\underbrace{\left\{
\begin{array}{c}
\sigma_x \\
\sigma_y \\
\sigma_{xy}
\end{array}
\right\}}_{\vec{\sigma}}
=
\underbrace{\frac{E}{1-v^2}
\begin{bmatrix}
1 & v & 0 \\
v & 1 & 0 \\
0 & 0 & \dfrac{1-v}{2}
\end{bmatrix}}_{D}
\underbrace{\left\{
\begin{array}{c}
\varepsilon_x \\
\varepsilon_y \\
\varepsilon_{xy}
\end{array}
\right\}}_{\vec{\varepsilon}}
\tag{4.25}
$$

The equation of balance represents the PDE for the Cauchy stress tensor σ_{ij}. The Cauchy stress tensor is a tensor consisting of nine components, where the origin is a certain point O and a micro rectangle made of three coordinate planes and three parallel planes around it, and the surface force (T_x, T_y, T_z) coordinate component acting on each plane perpendicular to the x, y, z axis is written as $\{(\sigma_x, \sigma_{xy}, \sigma_{xz}), (\sigma_{yx}, \sigma_y, \sigma_{yz}), (\sigma_{zy}, \sigma_{zy}, \sigma_z)\}$. This set of nine stress components is called the point stress or the Cauchy stress tensor. Here, σ_{ii} is the vertical stress and $\sigma_{ij(i\neq j)}$ is the shear stress, the index in σ_{ii} indicates the direction of action of the component (the same as the normal direction of the plane), the first index in $\sigma_{ij(i\neq j)}$ indicates the direction normal to the plane, and the second index the direction of action of the component. Also, f_i here represents the volumetric force. The volumetric force is an important element in the equation for the forces to which a structure is subjected. The volumetric force represents the force created by the load applied to a structure. Loads applied to a structure include gravity, wind pressure, and earthquakes. These loads are called volumetric forces because they act equally on the entire volume of the structure.

In the relationship between strain and displacement, $\vec{u}$ represents the displacement vector. $\frac{\partial u}{\partial x}$ and others represent their partial derivatives. $\vec{\varepsilon}$ is the strain vector.

In the constitutive equations, E and ν denote Young's modulus and Poisson's ratio. Using the equations of balance, the relationship between strain and displacement, and the constitutive equation, a second-order PDE for the displacement $\vec{u}$ can be obtained.

$$
\begin{cases}
\nabla^{*T}(D\nabla^*\vec{u}) + \vec{f} = 0 \\
\vec{u} = \bar{\vec{u}} \ on \ \Gamma_u \\
N_x D\nabla^*\vec{u} = \bar{\vec{t}} \ on \ \Gamma_t
\end{cases}
\tag{4.26}
$$

By writing this equation in matrix form and transforming it, we can obtain two partial differential equations regarding the displacement vector component $\begin{pmatrix} u \\ v \end{pmatrix}$.

$$
\frac{E}{1-\nu^2}
\left\{
\begin{pmatrix}
\dfrac{\partial}{\partial x} & 0 & \dfrac{\partial}{\partial y} \\[2mm]
0 & \dfrac{\partial}{\partial y} & \dfrac{\partial}{\partial x}
\end{pmatrix}
\begin{pmatrix}
1 & \nu & 0 \\
\nu & 1 & 0 \\
0 & 0 & \dfrac{1-\nu}{2}
\end{pmatrix}
\begin{pmatrix}
\dfrac{\partial}{\partial x} & 0 \\[2mm]
0 & \dfrac{\partial}{\partial y} \\[2mm]
\dfrac{\partial}{\partial y} & \dfrac{\partial}{\partial x}
\end{pmatrix}
\right\}
\vec{u} + \vec{f} = 0
$$

$$
\frac{E}{1-\nu^2}
\left\{
\begin{pmatrix}
\dfrac{\partial}{\partial x} & 0 & \dfrac{\partial}{\partial y} \\[2mm]
0 & \dfrac{\partial}{\partial y} & \dfrac{\partial}{\partial x}
\end{pmatrix}
\begin{pmatrix}
\dfrac{\partial}{\partial x} & \nu\dfrac{\partial}{\partial y} \\[2mm]
\nu\dfrac{\partial}{\partial x} & \dfrac{\partial}{\partial y} \\[2mm]
\dfrac{1-\nu}{2}\dfrac{\partial}{\partial y} & \dfrac{1-\nu}{2}\dfrac{\partial}{\partial x}
\end{pmatrix}
\right\}
\vec{u} + \vec{f} = 0
$$

$$
\frac{E}{1-\nu^2}
\left\{
\begin{pmatrix}
\dfrac{\partial^2}{\partial x^2} + \dfrac{1-\nu}{2}\dfrac{\partial^2}{\partial y^2} & \dfrac{1+\nu}{2}\dfrac{\partial^2}{\partial y\partial x} \\[3mm]
\dfrac{1+\nu}{2}\dfrac{\partial^2}{\partial y\partial x} & \dfrac{\partial^2}{\partial y^2} + \dfrac{1-\nu}{2}\dfrac{\partial^2}{\partial x^2}
\end{pmatrix}
\right\}
\begin{pmatrix} u \\ v \end{pmatrix}
+
\begin{pmatrix} f_x \\ f_y \end{pmatrix}
= 0
$$

$$
\frac{E}{1-\nu^2}\left(\frac{\partial^2 u}{\partial x^2} + \frac{1-\nu}{2}\frac{\partial^2 u}{\partial y^2} + \frac{1+\nu}{2}\frac{\partial^2 u}{\partial y\partial x}\right) + f_x = 0
$$

$$
\frac{E}{1-\nu^2}\left(\frac{\partial^2 v}{\partial y^2} + \frac{1-\nu}{2}\frac{\partial^2 v}{\partial x^2} + \frac{1+\nu}{2}\frac{\partial^2 v}{\partial y\partial x}\right) + f_y = 0
$$

For example, if Young's modulus is $200000\,\mathrm{N/mm}^2$ and Poisson's ratio is 0.3, we can write concrete partial differential equations. The body force is calculated from the point load applied to the boundary.

When the definition domain and boundary conditions are set as follows,

$$(x, y) \in [0, 1] \times [0, 1]$$

$$\begin{cases} \sigma_{xx}|_{x=0} = 0 \\ \sigma_{xx}|_{x=1} = 0 \\ \sigma_{yy}|_{y=1} = (\lambda + 2\mu)Q\sin(\pi x) \end{cases}$$

$$\begin{cases} v|_{x=0} = 0 \\ v|_{x=1} = 0 \\ u|_{y=0} = 0 \\ u|_{y=1} = 0 \end{cases}$$

the exact solution is obtained as follows: Here, μ and λ are Lamé constants defined by Young's modulus E and Poisson's ratio ν, which is computed as follows:

$$\lambda = \nu E/(1+\nu)(1-2\nu)$$

$$\mu = E/(2(1+\nu))$$

$$u = \cos(2\pi x)\sin(\pi y) \qquad (4.27)$$

$$v = \sin(\pi x)\,Qy^3/4 \qquad (4.28)$$

In the exact solution of the stress equation, the domain of definition of $x \in [0, 1]$ and $y \in [0, 1]$ is sampled, and $u(x, y), v(x, y)$ is computed based on their coordinates, which are considered pseudo-measurement data. The partial differential term candidate library is constructed as follows:

$$\Phi = [u_x, \mathbf{u_{xx}}, u_{xxx}, v_x, \mathbf{v_{xx}}, v_{xxx}, u_y, \mathbf{u_{yy}}, u_{yyy}, v_y, \mathbf{v_{yy}}, v_{yyy}]$$

The results are shown in the top color map diagram in Figure 4.27. First, derivation of the linear stress equation is performed in N10. Results

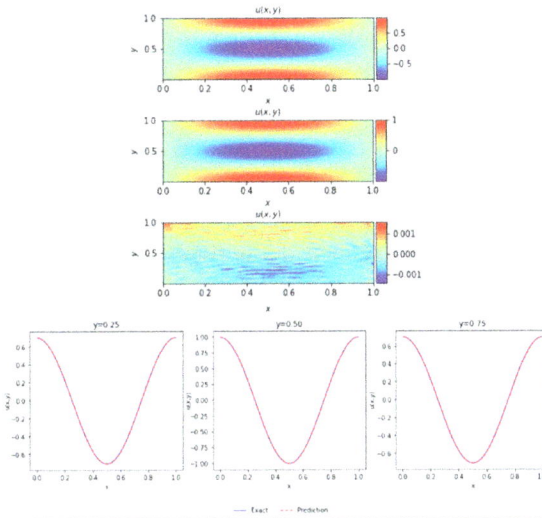

Correct PDE	$u_{xx} - 0.35u_{\Box} + 0.65u_{x} + 0.91f/E = 0$
Identified PDE (clean data)	$0.750783u_{\Box} - 0.001252u_{\Box} - 0.054714u + 1.333181u_{\Box} - 0.007121u_{\Box}$ $+ 0.648956u_{\Box} + 0.000348u_{\Box} + 0.005218u_{\Box} + 0.91f/E = 0$

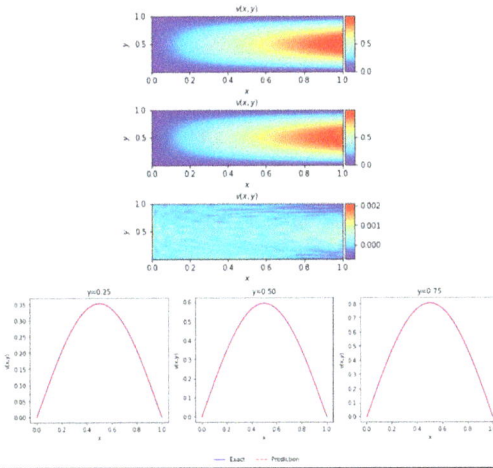

Correct PDE	$0.35v_{\Box} + v_{\Box} + 0.65v_{\Box} + 0.91f/E = 0$
Identified PDE (clean data)	$0.008128v_{\Box} + 0.574188v_{\Box} - 0.000791v_{\Box} + 2.175063v_{\Box} + 1.784790v_{\Box}$ $+ 0.007119v_{\Box} + 0.649728v_{\Box} - 0.009950v_{\Box} - 0.004686v_{\Box} + 0.91f/E = 0$

Figure 4.27. Results of two-dimensional stress equation derivation with extended library.

in which the NN has the minimum error are shown in the middle color map diagram in Figure 4.27. The lower color map diagram in Figure 4.27 shows the error distribution $(\hat{u} - u)$ and $(\hat{v} - v)$ between the NN model and the exact solution.

The data distribution $\{u(x, y) \mid y = 0.25, 0.5, 0.75\}$ at fixed time is also shown below the error distribution. In addition, the bottom row of Figure 4.27 shows the derived PDE, but the derivation error is not sufficient because of the large number of candidate partial derivative terms in the library.

The derivation error improves when the candidate partial differential terms in the library are minimized to only the exact ones. That is, if the partial differential term candidate library is configured as shown in $\Phi = [u_x, u_y, u_{xx}, u_{yy}, v_x, v_y, v_{xx}, v_{yy}]$, Figure 4.28 shows the exact solution, NN model, error distribution, data distribution at fixed time, and derived PDE.

Figures 4.29 and 4.30 show the parallel coordinates of the integrated error, E_{int}, when the approximate solution of the stress equation is computed in N10.

4.1.4.8. *Three-dimensional advection–diffusion equation*

As explained in Section 1, Section 4, the basic formula is as follows:

$$\frac{\partial u}{\partial t} = \nabla \cdot (D\nabla u) - \nabla \cdot (\vec{v}u) + R \tag{4.29}$$

$$\frac{\partial u}{\partial t} = D\left(\frac{\partial^2 u}{\partial x^2} + \frac{\partial^2 u}{\partial y^2} + \frac{\partial^2 u}{\partial z^2}\right) - v_x \frac{\partial u}{\partial x} - v_y \frac{\partial u}{\partial y} - v_z \frac{\partial u}{\partial z} + R$$

In the definition domain, $(t, x, y, z) \in [0, 1] \times [0, 1] \times [0, 1] \times [0, 1]$ when the following initial conditions $u|_{t=0} = \exp\left(-\frac{(x-C_x)^2}{\alpha_x} - \frac{(y-C_y)^2}{\alpha_y} - \frac{(z-C_z)^2}{\alpha_z}\right)$ are applied to it, the exact equation can be derived as follows:

$$u = \frac{1}{(4t+1)^{\frac{3}{2}}} \exp\left(-\frac{(x - v_x t - C_x)^2}{\alpha_x(4t+1)} - \frac{(y - v_y t - C_y)^2}{\alpha_y(4t+1)} - \frac{(z - v_z t - C_z)^2}{\alpha_z(4t+1)}\right) \tag{4.30}$$

In this exact solution, the definition domain of $t \in [0, 1]$, $x \in [0, 1]$, $y \in [0, 1]$, $z \in [0, 1]$ is sampled, and $u(t, x, y, z)$ is computed based on its coordinate values, which is considered pseudo-measurement data. The partial differential term candidate library is constructed as follows:

$$\Phi = [u_x, u_{xx}, u_{xxx}, u_y, u_{yy}, u_{yyy}, u_z, u_{zz}, u_{zzz}]$$

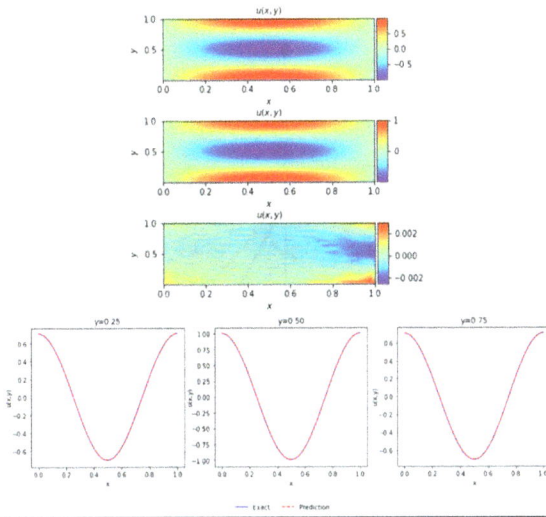

Correct PDE:	$u_{xx} + 0.35u_{yy} + 0.65u_{xy} + 0.9If/E = 0$
Identified PDE (clean data)	$0.947232u_{xx} + 0.563090u_{yy} + 0.648058u_{xy} + 0.9If/E = 0$

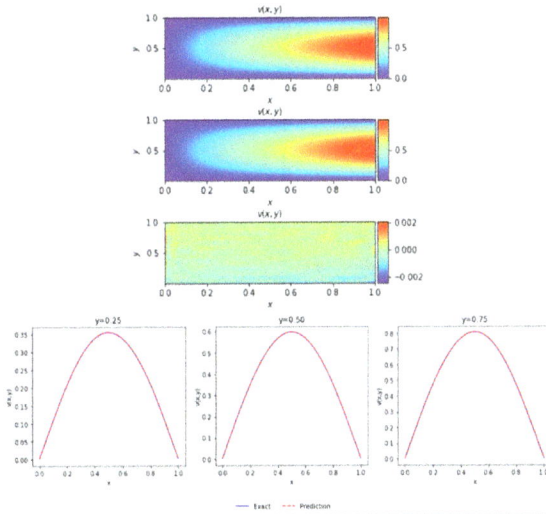

Correct PDE	$0.35v_{xx} + v_{yy} + 0.65v_x + 0.9If/E = 0$
Identified PDE (clean data)	$0.336581v_{xx} + 1.036389v_{yy} + 0.640864v_x + 0.9If/E = 0$

Figure 4.28. Results of two-dimensional stress equation derivation with minimum library.

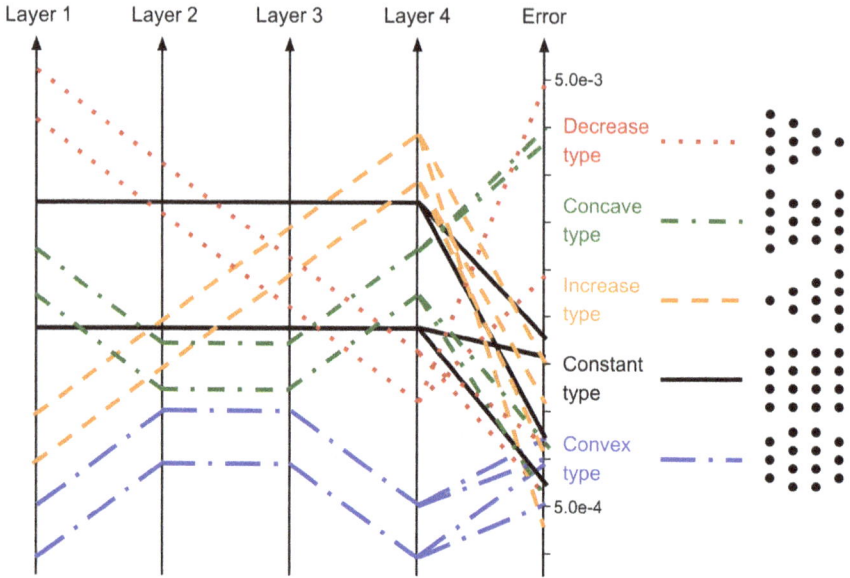

Figure 4.29. NN structure and E_{int} using extended library.

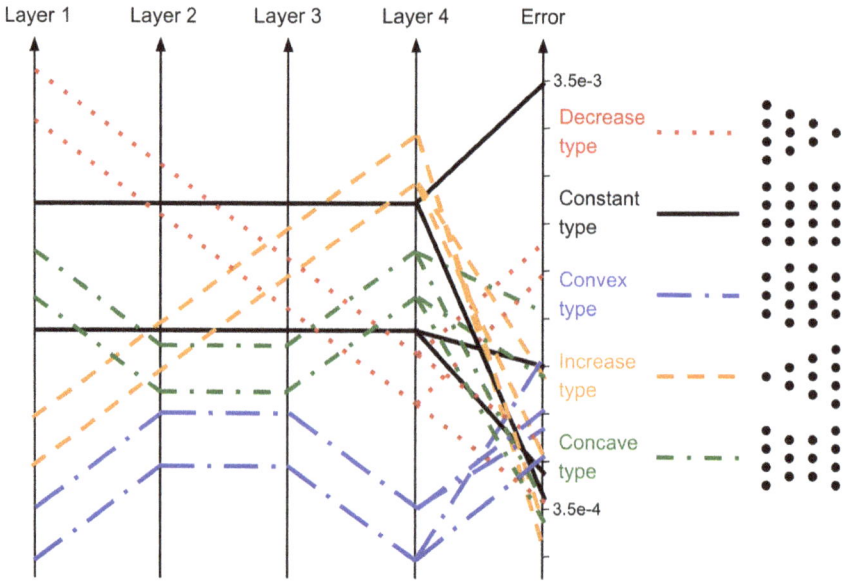

Figure 4.30. NN structure and E_{int} using minimum library.

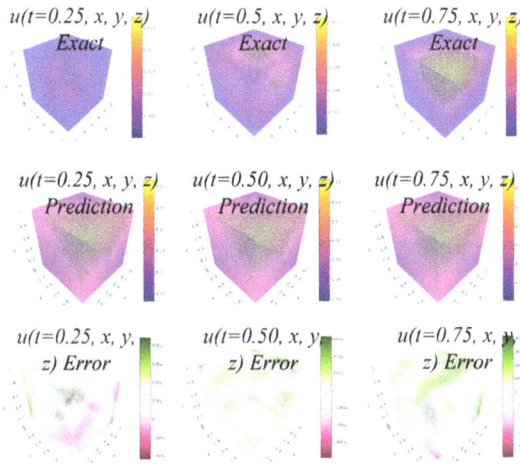

u(t=0.25, x, y, z) Exact	u(t=0.5, x, y, z) Exact	u(t=0.75, x, y, z) Exact
u(t=0.25, x, y, z) Prediction	u(t=0.50, x, y, z) Prediction	u(t=0.75, x, y, z) Prediction
u(t=0.25, x, y, z) Error	u(t=0.50, x, y, z) Error	u(t=0.75, x, y, z) Error

Correct PDE	$u_t - (u_{xx} + u_{yy} + u_{zz}) + (u_x + 1.1u_y + 1.2u_z)$ 0
Identified PDE (clean data)	$u_t - (0.855650u_{xx} + 0.957342u_{yy} + 0.937976u_{zz}) + 0.945638u_x + 1.269608u_y + 1.373395u_z + 0.025547u_{xy} - 0$

Figure 4.31. Results of three-dimensional advection–diffusion equation derivation with extended library.

The results are shown in the top color map diagram in Figure 4.31. As in the stress equation, an approximate solution to the advection–diffusion equation is first computed using a nine-layer NN with 20 neurons per hidden layer. The NN model is then constructed. The results are shown in the middle color map diagram in Figure 4.31. The lower color map diagram in Figure 4.31 shows the error distribution $(\hat{u} - u)$ between the NN model and the exact solution.

The data distribution $\{u(x, y, z) \mid t = 0.25, 0.5, 0.75\}$ at fixed time is also shown below the error distribution. In addition, the bottom row of Figure 4.31 shows the derived PDE, but the derivation error is not sufficient because of the large number of candidate partial derivative terms in the library.

The derivation error improves when the candidate partial differential terms in the library are minimized to only the exact ones. That is, if the partial differential term candidate library is configured as shown in $\Phi = [u_x, u_{xx}, u_y, u_{yy}, u_z, u_{zz}]$, Figure 4.32 shows the exact solution, NN model, error distribution, data distribution at fixed time, and derived PDE.

Figures 4.33 and 4.34 show the parallel coordinates of the integrated error, E_{int} when the approximate solution of the stress equation is computed in N10.

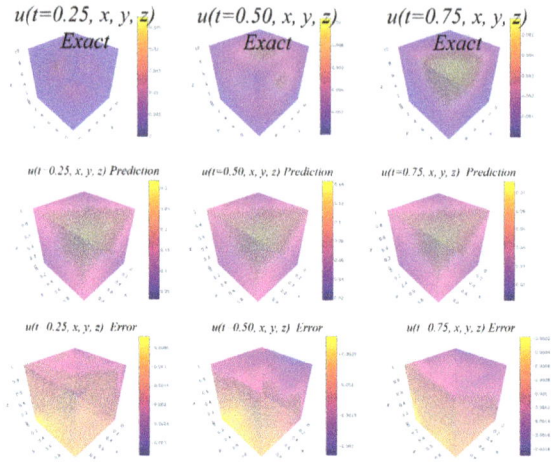

Correct PDE	$u_t - (u_{xx} + u_{yy} + u_{zz}) + (u_x + 1.1u_y + 1.2u_z) = 0$
Identified PDE (clean data)	$u_t - (0.866575u_{xx} + 0.967888u_{yy} + 0.917005u_{zz}) + 1.084462u_x + 1.235571u_y + 1.309882u_z = 0$

Figure 4.32. Results of three-dimensional advection–diffusion equation derivation with minimum library.

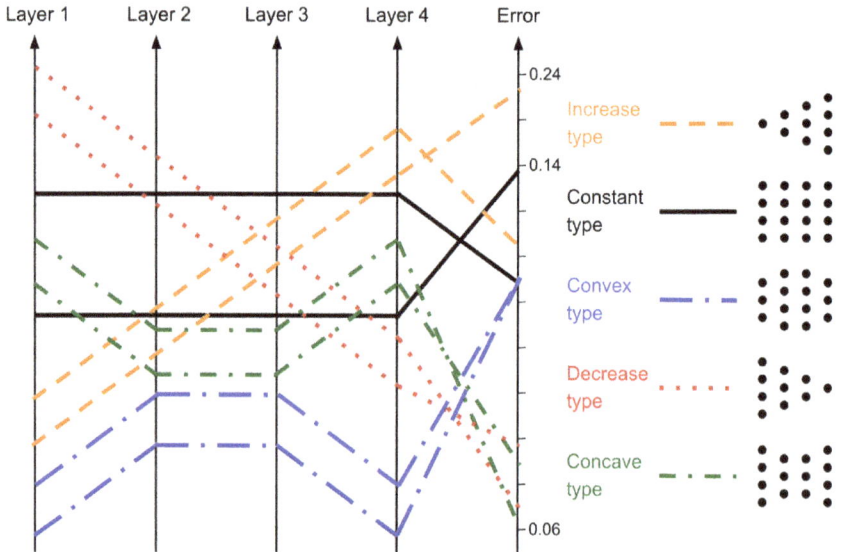

Figure 4.33. NN structure and E_{int} using extended library.

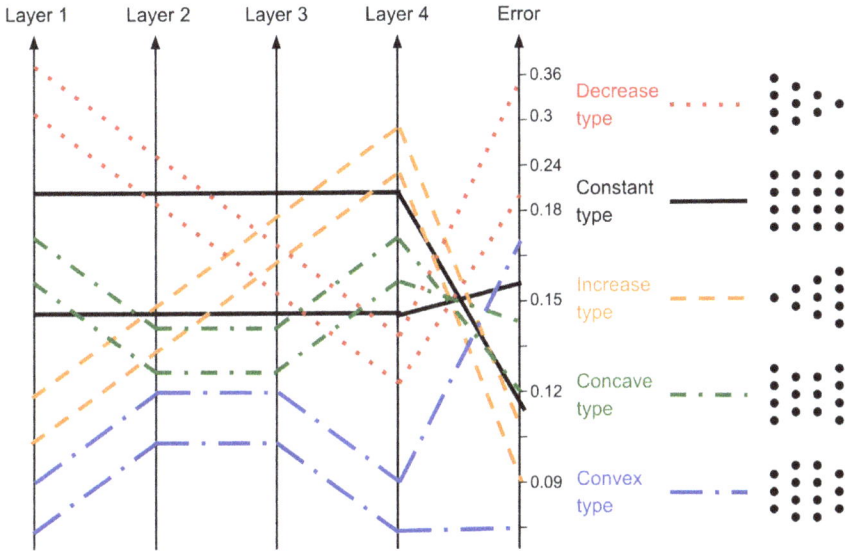

Figure 4.34. NN structure and E_{int} using minimum library.

4.2. Use of visual analysis techniques

This book has so far described methods for deriving PDEs and solving them from physical data using NNs. As discussed earlier, this book assumes that physical data consist of the space-time coordinates in which the data are defined and the physical data associated with those coordinates. In NN, the space-time coordinates are used as input and the physical data as output, and the function that approximates the physical data is regressed. Yet, the following two aspects are largely left to humans:

(1) What kind of NN to build
(2) How to construct a partial differential term library

Therefore, it is desirable to develop visual analytics (VA) methods that emphasize the importance of visual exploration alongside the process that automatically constructs models.

As for the accuracy of the results, a relationship with the NN hyperparameters was observed. However, increasing the complexity of the NN (number of layers, number of neurons per layer, etc.) does not necessarily increase the accuracy. Therefore, it is an important issue for the future to optimize NN hyperparameters. It is important to build hyperparameter visualization infrastructure that can utilize the judgment of users, in addition to simply using automation techniques.

In NN, the number of units per layer is variable. For example, AutoEncoders, the algorithm for dimensionality reduction using NN, is structured in such a way that the number of units is first gradually reduced from the input layer and then increased. As a visualization method to represent such parameters naturally, NN is represented by multidimensional coordinate points with an upper limit on the number of layers (dimensions). What is expected is the visualization method that performs dimensional reduction and maps error values in two- or three-dimensional space (Figure 4.35).

The number of units per layer is entered in each row of the cell matrix (error value in the last cell) in the bottom row of Figure 4.35. In the first line, the NN consists of two layers, indicating that 80 units are allocated in each layer. The user interactively selects point groups with low errors on the scatterplot (nine in Figure 4.35), which are visualized in

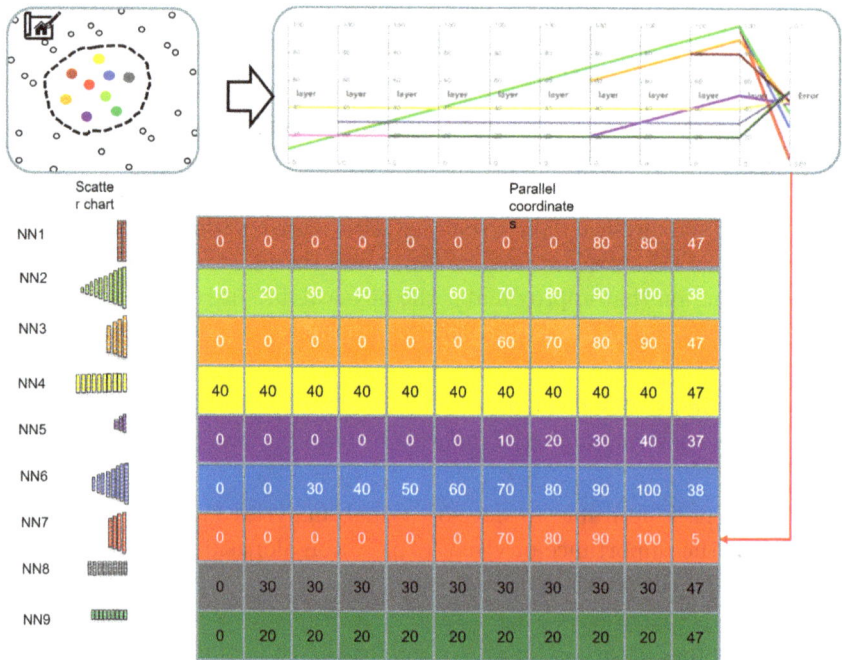

Figure 4.35. Visualization of NN hyperparameters.

parallel coordinates. Parallel coordinates assign each dimension of multidimensional data to a parallel coordinate axis and represent it as a polyline
connecting values for each dimension. The rightmost coordinate axis represents the error. This polyline is a coordinate point in a multidimensional
space. The NN label and its simplified representation are shown to the left
of the left end of the cell matrix. In the example shown in Figure 4.35, NN7
corresponding to the red point is selected. It indicates that for NN2 and
NN6, the errors are larger despite having the same pattern of the number
of units and the larger number of layers.

4.3. Methods for solving a given PDE

This book has so far described methods for deriving PDEs from physical data using NN models. As discussed earlier, this book assumes that
physical data consist of the space-time coordinates at which the data are
defined and the physical data associated with those coordinates. In NN,
the space-time coordinates are used as input and the physical data as output, and the function that approximates the physical data is regressed.
This book assumes PDEs with an exact solution and regards the exact
solution data of the PDE at multiple discrete space-time points as pseudo-
measurement data. As an application example, for a PDE with an exact
solution, this book discussed a method to find the original PDE with high
accuracy by considering the results computed by randomly changing the
space-time coordinates in the exact solution as the pseudo-measurement
data. The relationship between the NN hyperparameters (number of layers
and neurons) and the accuracy was then visualized.

This section discusses a method for solving a given PDE using NN with
boundary conditions (Dirichlet or Neumann type) described by space-time
discrete points. The Dirichlet boundary condition is a type of boundary
condition in PDE used when the value of a variable is given on the boundary.
For example, it is used to describe temperature distributions, potential
distributions, and fluid velocity distributions.

The Dirichlet boundary condition is mathematically expressed as follows: When a function u(x) is given on the boundary $\partial\Omega$ of some domain
Ω, it is expressed as

$$u(x) = g(x), \quad x \in \partial\Omega, \tag{4.31}$$

where g(x) represents the value on the boundary, and $u(x)$ the solution
inside the domain Ω.

For example, considering the two-dimensional heat conduction equation

$$\partial u/\partial t = \alpha \left(\frac{\partial^2 u}{\partial x^2} + \frac{\partial^2 u}{\partial y^2} \right),\tag{4.32}$$

when a temperature is given on a boundary within a rectangular region Ω using the Dirichlet boundary condition, it is expressed as follows:

$$u(x, y, t) = g(x, y), \ (x, y) \in \partial\Omega\tag{4.33}$$

where $g(x, y)$ represents the temperature on the boundary of region Ω and $u(x, y, t)$ represents the temperature within region Ω.

The Neumann boundary condition is a type of boundary condition in PDE used when the normal gradient of a variable is given on a boundary. For example, it is used to describe the thermal conductivity of an object surface or the flow rate of a fluid.

The Neumann boundary condition is mathematically expressed as follows: When a gradient in the normal direction of a function $u(x)$ is given on the boundary $\partial\Omega$ of a region Ω, it is expressed as

$$\left(\frac{\partial u}{\partial n} \right)(x) = g(x), \quad x \in \partial\Omega\tag{4.34}$$

where $\partial u/\partial n$ represents the gradient in the normal direction u and $g(x)$ the value on the boundary.

For example, considering the two-dimensional heat conduction equation

$$\frac{\partial u}{\partial t} = \alpha \left(\frac{\partial^2 u}{\partial x^2} + \frac{\partial^2 u}{\partial y^2} \right),\tag{4.35}$$

when a heat flux is given on the upper boundary of a rectangular region Ω using the Neumann boundary condition, it is expressed as follows:

$$(\partial u/\partial y)(x, y, t) = g(x, y), (x, y) \in \partial\Omega$$

where $g(x, y)$ represents the heat flux on the upper boundary of the region Ω, and $\partial u/\partial y$ represents the gradient of the temperature $u(x, y, t)$ in the region Ω in the y direction.

As physical data, they are defined at the boundaries and inside the region of interest, from which NN model errors and PDE errors are computed to obtain a solution of a PDE (Figure 4.37). NN model errors were explained in the previous section, and PDE errors will be explained later. As an application example, this section introduces a program that utilizes this method to find solutions for PDEs with known exact solutions and compares them with the exact results. Additionally, for PDEs lacking

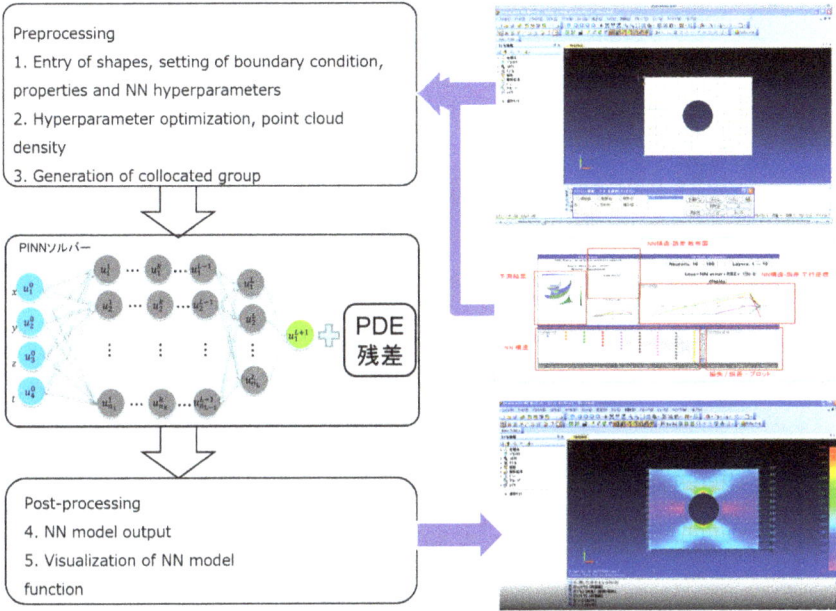

Figure 4.36. Detail of PINN processing and the user interface.

exact solutions, we compare them with conventional PDE approximation solutions.

4.3.1. *How to solve PDEs using the Fourier transform*

Before describing how to solve PDEs using NNs, this section introduces how to solve PDEs using the Fourier transform [42]. The following equation is used as a definition. Here, k is any real number.

$$f(x) = \frac{1}{\sqrt{2\pi}} \int_{-\infty}^{\infty} F(k)e^{ikx} dk \qquad (4.36)$$

If the explanatory variable x in the function of interest $f(x)$ is a physical quantity, the Fourier transform transfers the dimension of the explanatory variable to its original reciprocal. For example, when the explanatory variable x in the function before the transformation has the dimension of time, the explanatory variable k after the transformation has the dimension of frequency. Alternatively, when the explanatory variable x before the transformation has a length dimension, the explanatory variable k after the transformation has a wavenumber dimension. Based on the definition, this property holds from the fact that kx is a dimensionless quantity.

$$F(k) = \frac{1}{\sqrt{2\pi}} \int_{-\infty}^{\infty} f(k)e^{-ikx}\,dx \qquad (4.37)$$

In general, $f(x)$ is a function that decays to zero at the endpoints of the domain of definition. The Fourier transform is used when focusing on spatial distribution. By the way, the Laplace transform is essentially the same as the Fourier transform.

$$\begin{cases} f(t) = \dfrac{1}{2\pi i} \displaystyle\int_{a-i\infty}^{a+i\infty} F(p)e^{pt}\,dp & (4.38) \\[3mm] F(p) = \displaystyle\int_{-\infty}^{\infty} f(t)e^{-pt}\,dt & (4.39) \end{cases}$$

Now, let us solve the following diffusion equation (domain of definition: infinite) using the Fourier transform.

$$\frac{\partial}{\partial t}u(x,t) = \frac{\partial^2}{\partial x^2}u(x,t) \qquad (4.40)$$

$$u(x,0) = f(x)$$

The spatial distribution provides a hint, so we apply the Fourier transform to both sides. The $u(x,t)$ is a bivariate function, but this time we will perform the Fourier transform on x.

$$U(k,t) = \frac{1}{\sqrt{2\pi}} \int_{-\infty}^{\infty} u(x,t)e^{-ikx}\,dx \qquad (4.41)$$

The left side is $\frac{\partial}{\partial t}U(k,t)$

The right side is transformed using the partial integration method.

$$\frac{1}{\sqrt{2\pi}} \int_{-\infty}^{\infty} \frac{\partial u}{\partial t}e^{-ikx}\,dx = \frac{1}{\sqrt{2\pi}}[u(x,t)e^{-ikx}]_{-\infty}^{\infty} + \frac{ik}{\sqrt{2\pi}} \overbrace{\int_{-\infty}^{\infty} u(x,t)e^{-ikx}\,dx}^{U(k,t)}$$

$$(4.42)$$

Therefore, $\frac{\partial}{\partial t}U(k,t) = (ik)^2 U(k,t)$

$$U(k,t) = Ae^{-k^2 t}$$

$U(k,0) = \frac{1}{\sqrt{2\pi}} \int_{-\infty}^{\infty} f(k)e^{-ikx}\,dx$ this means $U(k,0) = F(k)$

Therefore,

$$U(k,t) = F(k)e^{-k^2 t}$$

$$u(x,t) = \frac{1}{\sqrt{2\pi}} \int_{-\infty}^{\infty} U(k,t)e^{ikx} dk$$

$$= \frac{1}{\sqrt{2\pi}} \int_{-\infty}^{\infty} F(k)e^{-k^2 t + ikx} dk$$

$$= \frac{1}{2\pi} \int_{-\infty}^{\infty} dk \int_{-\infty}^{\infty} dx' f(x')e^{-k^2 t + ikx - ikx'}$$

We now apply the Gaussian integral

$$\int_{-\infty}^{\infty} e^{-\frac{a}{2}x^2} dx = \sqrt{\frac{2\pi}{a}} \qquad (4.43)$$

to the integral with respect to k.

The power part of the exponential function becomes as follows:

$$-k^2 t + ik(x - x') = -t\left(k - \frac{i(x - x')}{2t}\right)^2 - \frac{(x - x')^2}{4t}$$

$$= \frac{1}{2\pi}\sqrt{\frac{\pi}{t}} \int_{-\infty}^{\infty} dx' f(x')e^{-\frac{(x-x')^2}{4t}}$$

If $f(x) = \delta(x)$, then $\int_{-\infty}^{\infty} dx' \delta(x')e^{-\frac{(x-x')^2}{4t}} = e^{-\frac{x^2}{4t}}$, so we get the following general solution:

$$u(x,t) = \frac{1}{2\sqrt{\pi t}}e^{-\frac{x^2}{4t}} \qquad (4.44)$$

Here, $\delta(x)$ is called Dirac's delta function which becomes 1 when integrated.

4.3.2. *PDE approximate solution*

Methods to obtain the approximate solution of PDE include FDM, FEM, and particle method (PM). In FDM, the approximate solution is obtained by replacing the differential with the difference [43]. In contrast, in FEM, an approximate solution is obtained by expanding an unknown function into a finite sequence of known functions and determining the unknown constants introduced in the expansion [28]. In FEM, the weighted residual method is used to determine the unknown constants. PM is one of the discretization methods for numerically solving PDEs on continua, where the object to be computed is represented as a collection of particles [44]. It is mainly used for fluid and structural analyses. In particular, in fluid analysis, it belongs to the Lagrange multiplier, which has the feature of eliminating the need to calculate the convection term. The particle method sets a parameter called radius of influence. The radius of influence refers to the radius used to find the local physical quantity (density, gradient value of pressure, Laplacian of velocity, etc.) at a given location. It is often used to count the number of particles within this radius or to take a weighted average of the physical quantity.

The method to solve PDEs using NNs described here is similar to PM in that it does not require a difference grid or finite elements. But it differs significantly in that it does not require the radius of influence to be set up to compute the partial differential terms. It shares some common points with FEM in that it evaluates PDE residuals at points within the region of interest. By explaining the weighted residuals that characterize FEM, its difference from the method to solve PDEs using NNs is clarified.

To explain the weighted residual method, this section examines the steady-state heat conduction problem in the two-dimensional analytical domain v as shown in Figure 4.37. Now, the problem in this case is defined by the following equation [27]: That is, governing equation: $\kappa(u_{xx} + u_{yy}) + Q = 0$ (in v) is solved under the boundary condition: $u = 0$ (on s) where u is the temperature, κ thermal conductivity, and Q heat generation rate per unit volume.

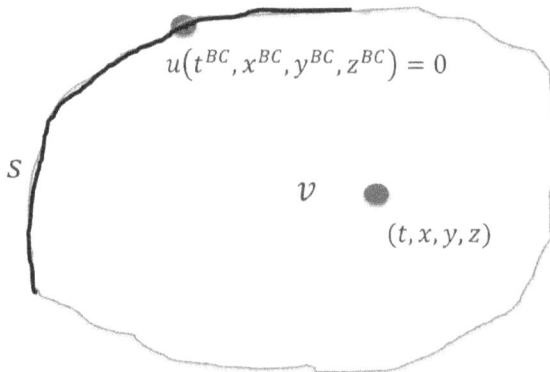

$$u\left(t^{BC}, x^{BC}, y^{BC}, z^{BC}\right) = 0$$

v

(t, x, y, z)

S

Figure 4.37. Domain to find solution to PDE.

Now, in the domain v, the unknown function u is approximated using the known sequence of functions u_j, $j = 1, N$ as follows:

$$u = \sum_{j=1}^{n} a_j u_j \qquad (4.45)$$

Here, be sure to select u_j, $j = 1, N$ that becomes zero on the boundary s.

The u_j is called the trial function, and a_j is an unknown constant. Equation (4.45) is substituted into the governing equation to obtain the residual

$$R(x, y) = \kappa \left(\sum_{j=1}^{n} a_j (u_{j_{xx}} + u_{j_{yy}}) \right) + Q. \qquad (4.46)$$

If the test function is an exact solution to the steady-state heat conduction problem, then the residual R should be zero. However, when the exact solution is not known and an approximate solution is to be obtained, the residual cannot be zero. Therefore, the weighted residual method determines a_j so that the residuals become zero in the average sense. That is, the residual R is multiplied by some weighting function w_i and integrated, which is set as zero. This is expressed in the equation as

$$\iint_v w_i R \, dv = 0, \quad i = 1, n. \qquad (4.47)$$

If n different functions are used in this equation as w_i, equation (4.47) becomes a series of n linear equations. Various weight functions w_i can be chosen, and two typical ones are shown in the following:

- Collocation method

The Dirac delta function is chosen as the weight function. That is,

$$w_i(\boldsymbol{x}) = \delta(\boldsymbol{x} - \boldsymbol{x_i}). \qquad (4.48)$$

However, $\boldsymbol{x}$ is a coordinate vector. This function has the property of

$$\iint_v \delta(\boldsymbol{x} - \boldsymbol{x_i}) R \, dv = R(\boldsymbol{x_i}). \qquad (4.49)$$

Thus, in this method, the residual R is determined so that the residual becomes zero at the specified n collocation points. As n increases, the residuals become zero at many points, and the entire domain is expected to eventually become zero.

The method to find a solution to PDE using NN can be classified as a collocation method. Here, in domain v, in the NN model, the unknown function $u(t, x, y, z)$ is approximated using the known activation function sequence $\sigma_{i_{j+1}}^j(x)$, $j = 0, L$ as follows:

$$u(t, x, y, z) = u_1^{L+1} = \sigma_1^L \left(\sum_{i_L=1}^{n_L} w_{i_L, i_{L+1}}^L \sigma_{i_L}^{L-1} \left(\sum_{i_{L-1}=1}^{n_{L-1}} w_{i_{L-1}, i_L}^{L-1} \cdots \right. \right.$$

$$\left. \left. \times \sigma_{i_1}^0 \left(\sum_{i_0=1}^{n_0} w_{i_0, i_1}^0 u_{i_0}^0 \right) \cdots \right) \right) \tag{4.50}$$

The n_0 represents the number of explanatory variables, which is $\begin{aligned} t &= u_1^0 \\ x &= u_2^0 \\ y &= u_3^0 \\ z &= u_4^0 \end{aligned}$.

In order to construct the NN model, the weights $w_{i_j, i_{j+1}}^j$, $j = 0, L$ are determined so that the residuals become zero in the average sense at points inside the target domain and on the boundary. This function is appropriately named the global collocation method (GCM) because, unlike FEM, it is determined for the whole domain at once.

4.3.2.1. *Finite element method (FEM)*

The Galerkin method, one of the key techniques of FEM, is the best known of the weighted residuals and was proposed by the Russian engineer Galerkin. In this method, the weight function w_i is made the same as the trial function T_i, that is, $w_i = T_i$. This method is closely related to the variational direct method and in that sense is important in FEM.

In order to obtain an approximate solution to the PDE, it is important to know how the evaluation of the partial differential term is realized. FDM employs a method of replacing partial derivatives with differences, as already discussed, and basically uses information from adjacent grids to compute the differences. To further improve accuracy, not only information from adjacent grids but also information from beyond adjacent grids is used. In FEM, the degree of the test function is increased in the elements to improve the approximation capability. In the particle method, the radius of influence is increased to improve the approximation capability. These methods are local interpolation. In contrast, GCM is a global approximation.

The sample code to perform the finite element solution method of the steady-state heat transfer equation in Colab is shown in the following:

```
< Start >
!pip install fenics
from fenics import *

# Create a rectangular mesh
mesh = UnitSquareMesh(8, 8)

# Define function space
V = FunctionSpace(mesh, "P", 1)

# Define boundary condition
u_D = Expression("1 + x[0]*x[0] + 2*x[1]*x[1]", degree=2)

def boundary(x, on_boundary):
    return on_boundary

bc = DirichletBC(V, u_D, boundary)

# Define variational problem
u = TrialFunction(V)
v = TestFunction(V)
f = Constant(-6.0)
a = dot(grad(u), grad(v))*dx
L = f*v*dx

# Compute solution
u = Function(V)
solve(a == L, u, bc)

# Plot solution
plot(u)
plt.show()
< End >
```

The code installs FEniCS, a finite element library, creates a unit square mesh, applies the FEM described later to it to obtain a solution, and plots the results. This code solves the two-dimensional Poisson's equation, and it can also solve three-dimensional equation and so forth by changing the number of dimensions.

4.3.2.2. *Finite difference method (FDM)*

This method converts the PDE into a difference form and solves it numerically. The difference form approximates the difference used to evaluate the solution of the PDE to a continuous function. It is usually expressed as a difference equation. PDEs solved using the difference method are discretized in both space and time. FDM solves PDEs using a finite difference approximation for discretized space and time. Note that FDM is also used to solve higher-order PDEs, but with limited accuracy.

The sample code for solving a heat transfer problem in Colab is shown in the following. This code is intended to numerically solve unsteady one-dimensional heat transfer problems using specific boundary conditions and thermal conductivities.

```
< Start >
import numpy as np
# As a boundary condition, set a temperature of 100°C at the left end and
    0°C at the right end.
T_left = 100
T_right = 0

# Set the thermal conductivity.
k = 0.1

# Set the space pitch width.
dx = 0.1

# Set the time pitch width.
dt = 0.01

# Set the initial temperature value.
T = np.zeros((101, 101))

# Compute time evolution.
for t in range(1, 10000):
    for i in range(1, 100):
        T[i, t] = T[i, t-1] + k * dt/dx**2 * (T[i+1, t-1] - 2*T[i, t-1] + T[i-1, t-1])
```

Apply boundary conditions.
T[0, t] = T_left
T[100, t] = T_right
< End >

This code divides space and time by a pitch width and computes their temperatures. FDM is used for space, and the Euler method is used for time to obtain temperature. The results of these computations can be used to obtain approximate solutions to steady-state heat transfer problems.

4.3.3. *How to find solutions to PDEs using PINNs*

In the methods to find solutions to PDEs discussed in this book, approximate solutions are computed while constructing an NN model from physical data representing Dirichlet boundary conditions and minimizing PDE residuals at the same time (Figure 4.38). Similar to the PDE derivation method, this method to find solutions to PDEs performs a multi-purpose parameter optimization computation. One of the purposes is to construct an NN model from physical data representing Dirichlet boundary conditions, and the parameter to be optimized is the NN weights. In other words, it searches for parameters that minimize the loss function. Another objective is to enhance the loss function by incorporating, through automatic differentiation, the residuals computed using the partial derivative terms of the NN model, in addition to the mean squared error. In addition, when a Neumann boundary condition is given, automatic differentiation is used to compute this boundary condition and its averaged squared error is added to the loss function. Figure 4.38 shows the case in which only the Dirichlet boundary condition is given.

From the Dirichlet boundary condition, the space-time coordinates in which the physical data are defined are taken out and entered into the input layer. The NN model error is computed by averaging the squared error between the data regressed by the NN and the physical data with the boundary point set to be trained. The PDE error is computed by averaging the PDE residuals given inside the domain of interest with the sampled point group. The sum of these two errors is used as a loss function to optimize the NN weight parameters. AD is used to compute the PDE residuals.

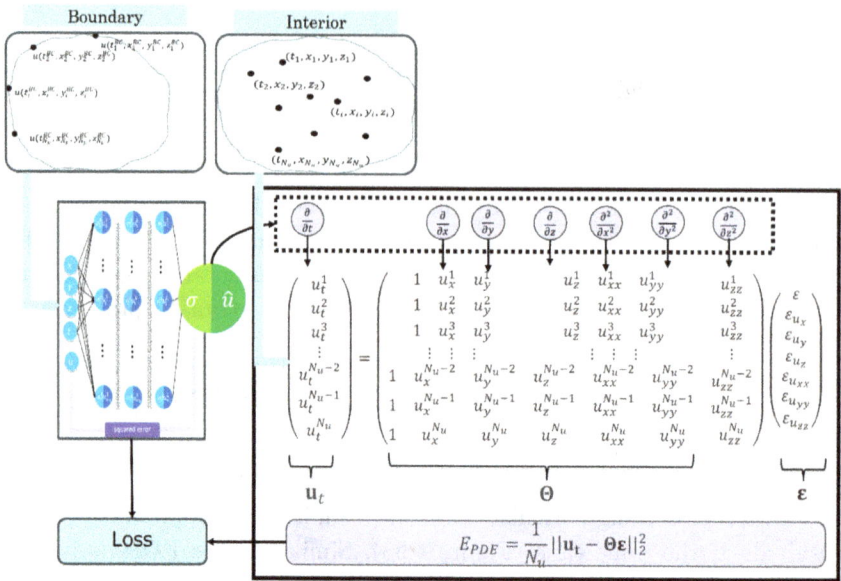

Figure 4.38. How to find solutions to PDE.

4.3.3.1. *Constructing a model using NNs*

Building an NN model involves optimizing NN parameters to minimize a loss function that quantifies the difference between the data specified by the Dirichlet boundary conditions and the data approximated by the NN model. The input is the coordinates (x, y, z, t) in space-time that are sampled at the boundary where the condition is given, and the output is the physical data u given at that point, such as electric potential, temperature, and pressure. In this case, the loss function consists only of the mean squared error $\text{MSE}_u = \|\hat{u} - u\|_2^2 / N_p$ between the physical data at the boundary and the data regressed by the NN model (hereinafter "the NN model error"). The circle labeled "σ" represents a neuron in each layer of the NN, and its activation function is σ. Also, N_p represents the number of physical data at the boundary.

4.3.3.2. *Computation of PDE residuals*

Using the previously described automatic differentiation (AD), the partial differential terms are computed from the NN model. In this course, NN models are built using Google TensorFlow. TensorFlow is a platform for using machine learning methods such as NN and optimization techniques. Here, the GradientTape library in TensorFlow is used to compute the partial differential terms of the NN model.

Next, the partial differential terms that make up the PDE are computed. For the NN model, the PDE is constructed using AD. If, as PDE, the first-order time partial derivative is expressed as a linear sum of spatial partial derivative terms, the regression equation in the following first expresses the relationship in N_u space-time coordinates sampled in the domain in which a solution is to be found:

$$u_t = \Theta \varepsilon \qquad (4.51)$$

where u_t is a vector consisting of first-order time partial derivative values, Θ is a matrix consisting of spatial partial derivative values at each level, and ε is the coefficient of their spatial partial derivative values. u_t, Θ is the number of measured points, that is, the number of space-time coordinates N_P. Unlike the PDE derivation, ε is a constant, not an unknown number. The vector consisting of the residuals R computed at each coordinate point is

$$R = u_t - \Theta \varepsilon. \qquad (4.52)$$

Next, the PDE error (MSE$_{\text{PDE}}$) is computed from this residual vector.

$$E_{\text{PDE}} = \frac{1}{N_P} \|R\|_2^2 = \frac{1}{N_P} \|u_t - \Theta \varepsilon\|_2^2 \qquad (4.53)$$

To evaluate a method to find a solution to PDE, it is important to define the error. As errors, the NN model errors (E_u) and PDE errors (E_{PDE}) described previously can be used. We will define the integrated error $E_{\text{int}} = E_U^* E_{\text{PDE}}$ as in the PDE derivation.

4.3.4. *Example of finding solutions to PDEs using PINNs*

4.3.4.1. *One-dimensional advection–diffusion equation*

This section compares the results of solving a one-dimensional advection–diffusion equation using PINNs with the results of solving it using FDM based on their exact solutions. For comparison, in FDM, Δx, Δt (mesh width) are determined so that the Courant number is 1 or less and the average is as large as possible within the domain. Figure 4.39 shows the results of comparison with FDM. Of these two methods, the PINN method provides the most accurate solution (Figure 4.39).

Results	Exact solution	FDM solution	PINN solution
u_t			
Errors	0	2.778628e-2	2.352048e-1
u_x			
Errors	0	2.515527e-1	3.859644e-1

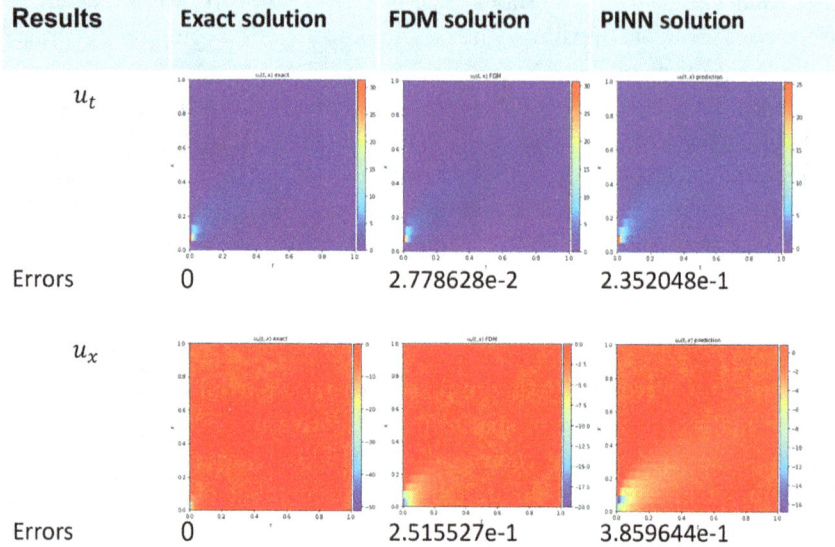

Figure 4.39. Comparative results of FDM and PINN methods in a one-dimensional advection–diffusion equation.

Figure 4.40 shows the parallel coordinates of E_{int} when the approximate solution of the AD equation is computed in N10.

4.3.4.2. *One-dimensional Burgers' equation*

This section compares the results of solving a one-dimensional Burgers' equation using PINNs with the results of solving it using FDM based on their exact solutions. For comparison, in FDM, Δx, Δt (mesh width) are determined so that the Courant number is 1 or less and the average is as large as possible within the domain. Figure 4.41 shows the results of comparison with FDM. Of these two methods, the PINN method provides the most accurate solution (Figure 4.41).

Figure 4.42 shows the parallel coordinates of E_{int} when the approximate solution of Burgers' equation is computed in N10.

4.3.4.3. *One-dimensional KdV equation*

This section compares the results of solving a one-dimensional KdV equation using PINNs with the results of solving it using FDM based on their

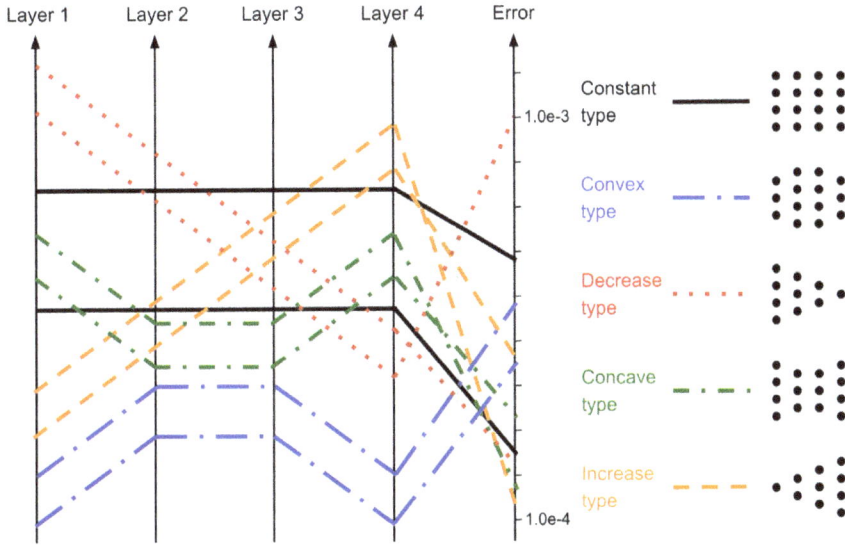

Figure 4.40. NN structure and E_{int}.

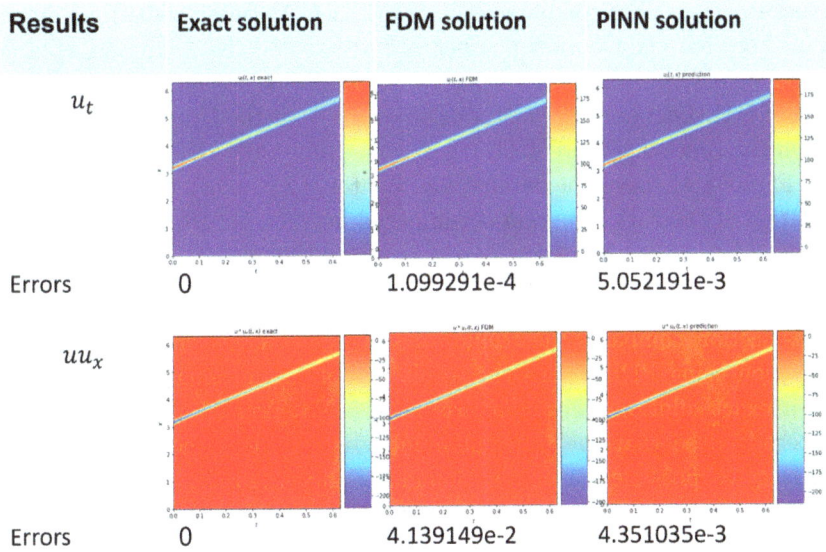

Figure 4.41. Comparison of FDM and PINN solution in a one-dimensional Burgers' equation.

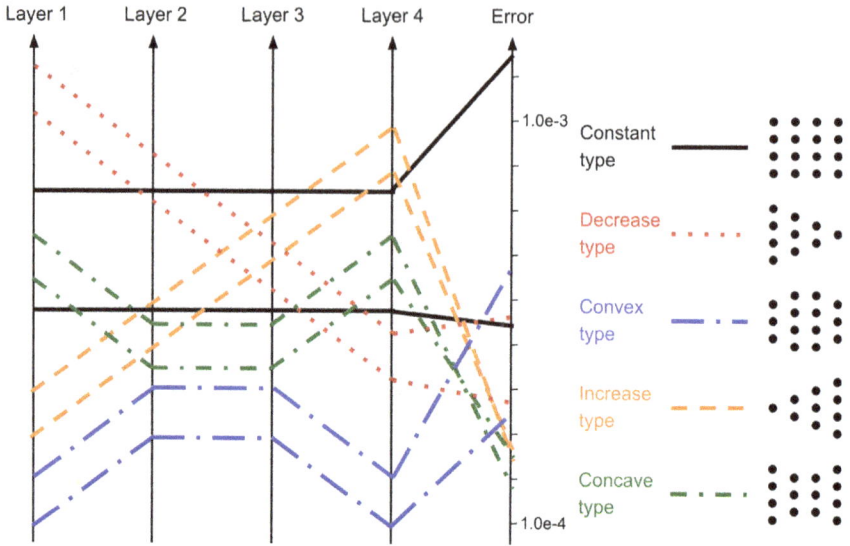

Figure 4.42. NN structure and E_{int}.

exact solutions. For comparison, in FDM, Δx, Δt (mesh width) are determined so that the Courant number is 1 or less and the average is as large
as possible within the domain. Figure 4.43 shows the results of comparison
with FDM. Of these two methods, the PINN method provides the most
accurate solution (Figure 4.43).

Figure 4.44 shows the parallel coordinates of E_{int} when the approximate
solution of the KdV equation is computed in N10.

4.3.4.4. *One-dimensional Poisson's equation*

This section compares the results of solving a one-dimensional Poisson's
equation using PINNs with the results of solving it using FDM based on
their exact solutions. For comparison, in FDM, Δx, Δt (mesh width) are
determined so that the Courant number is 1 or less and the average is
as large as possible within the domain. Figure 4.45 shows the results of
comparison with FDM. Of these two methods, the PINN method provides
the most accurate solution (Figure 4.45).

Figure 4.46 shows the parallel coordinates of E_{int} when the approximate
solution of Poisson's equation is computed in N10.

Results	Exact solution	FDM solution	PINN solution

u_t

| Errors | 0 | 1.372211e-3 | 1.108102e-3 |

uu_x

| Errors | 0 | 5.677015e-3 | 1.281659e-3 |

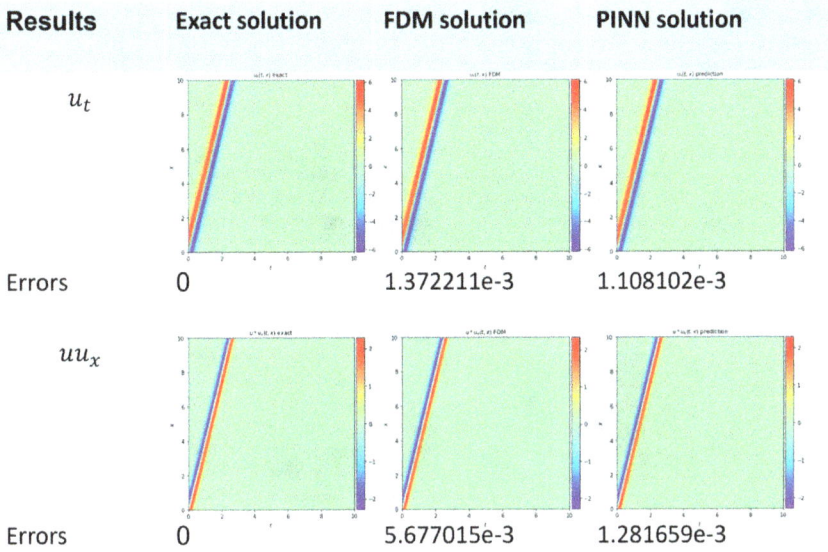

Figure 4.43. Comparison of methods to find solutions to the one-dimensional KdV equation using FDM and PINNs.

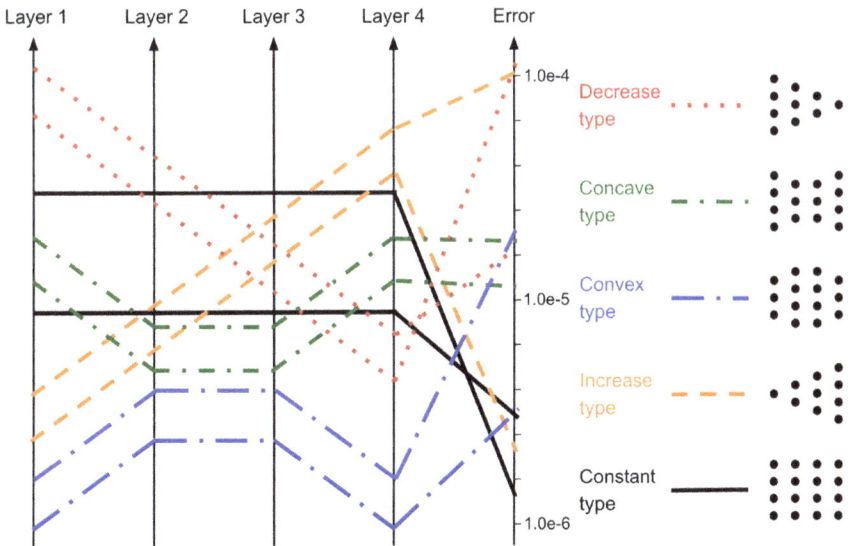

Figure 4.44. NN structure and E_{int}.

Results	Exact solution	FDM solution	PINN solution
u_t			
Errors	0	7.842592e-16	8.735365e-3
u_x			
Errors	0	4.188539e-2	1.187077e-2

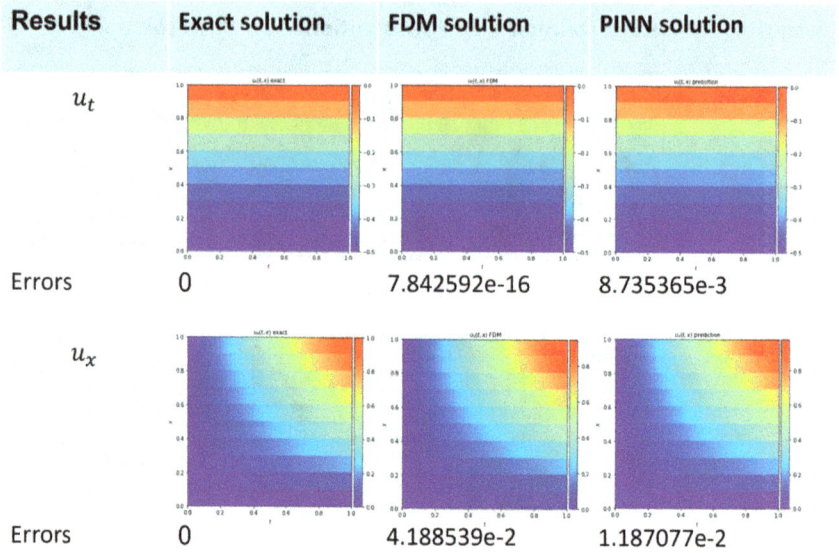

Figure 4.45. Comparison of FDM and PINN solution in a one-dimensional Poisson's equation.

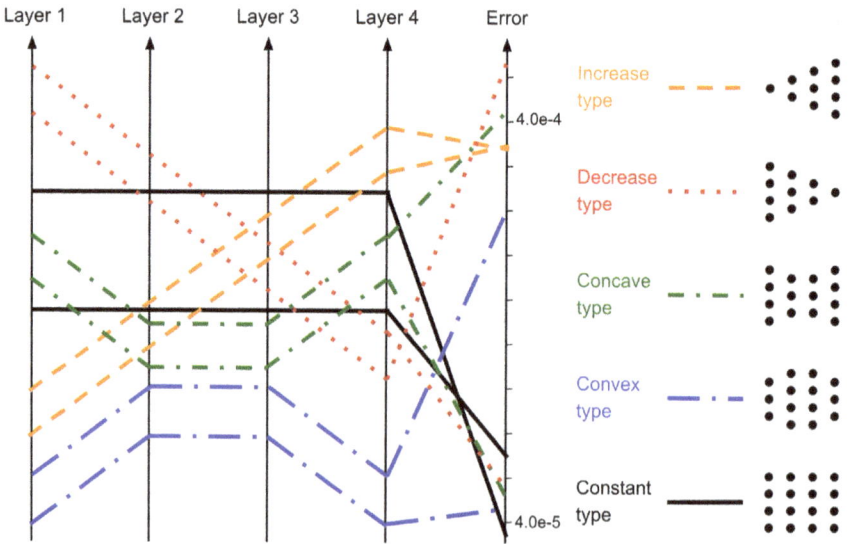

Figure 4.46. NN structure and E_{int}.

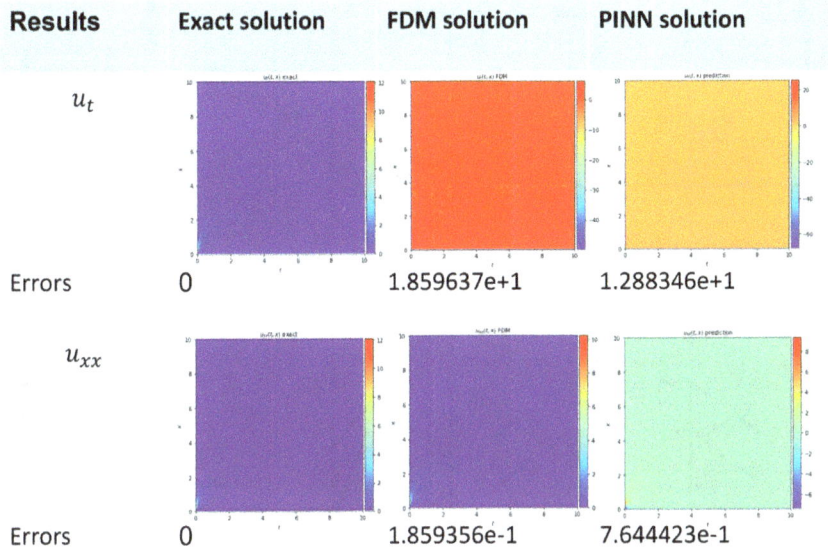

Results	Exact solution	FDM solution	PINN solution
u_t			
Errors	0	1.859637e+1	1.288346e+1
u_{xx}			
Errors	0	1.859356e-1	7.644423e-1

Figure 4.47. Comparison of FDM and PINN solution in a one-dimensional heat conduction equation.

4.3.4.5. *One-dimensional heat conduction problem*

This section compares the results of solving a one-dimensional heat conduction equation using PINNs with the results of solving it using FDM based on their exact solutions. For comparison, in FDM, Δx, Δt (mesh width) are determined so that the Courant number is 1 or less and the average is as large as possible within the domain. Figure 4.47 shows the results of comparison with FDM. Of these two methods, the PINN method provides the most accurate solution (Figure 4.47).

Figure 4.46 shows the parallel coordinates of E_{int} when the approximate solution of the heat conduction equation is computed in N10.

4.3.4.6. *One-dimensional wave equation*

This section compares the results of solving a one-dimensional wave motion equation using PINNs with the results of solving it using FDM based on their exact solutions. For comparison, in FDM, Δx, Δt (mesh width) are determined so that the Courant number is 1 or less and the average is as large as possible within the domain. Figure 4.48 shows the results of

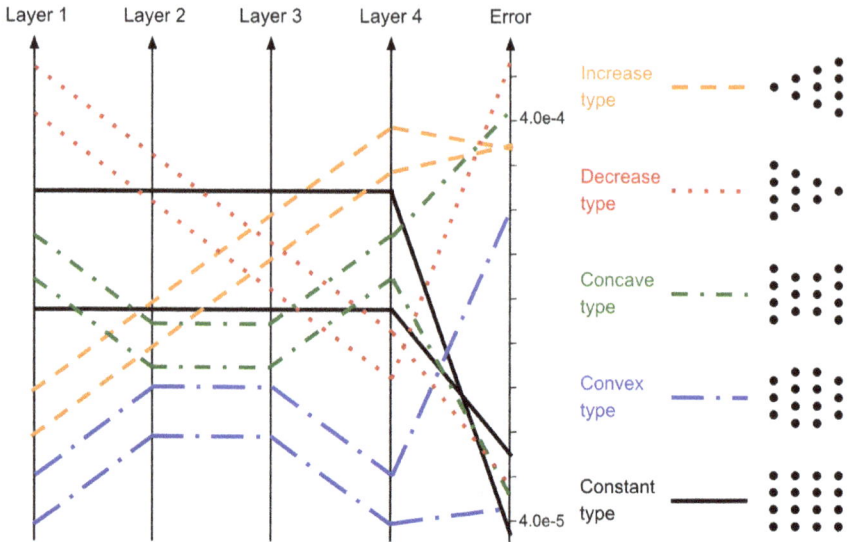

Figure 4.48. NN structure and E_{int}.

comparison with FDM. Of these two methods, the PINN method provides the most accurate solution (Figure 4.49).

Figure 4.50 shows the parallel coordinates of E_{int} when the approximate solution of the wave motion equation is computed in N10.

4.3.4.7. *Two-dimensional linear stress analysis*

This section compares the results of solving a two-dimensional stress equation using PINNs with the results of solving it using FDM based on their exact solutions. For comparison, in FDM, Δx, Δt (mesh width) are determined so that the Courant number is 1 or less and the average is as large as possible within the domain. Figure 4.51 shows the results of comparison with FDM. Of these two methods, the PINN method provides the most accurate solution (Figure 4.51).

Figure 4.52 shows the parallel coordinates of E_{int} when the approximate solution of the stress equation is computed in N10.

4.3.4.8. *Three-dimensional advection–diffusion equation*

This section compares the results of solving a three-dimensional advection–diffusion equation using PINNs with the results of solving it using FDM

Results	Exact solution	FDM solution	PINN solution
u_{tt}			
Errors	0	2.092094e-2	1.347749e-1
u_{xx}			
Errors	0	1.824482e-3	7.517304e-2

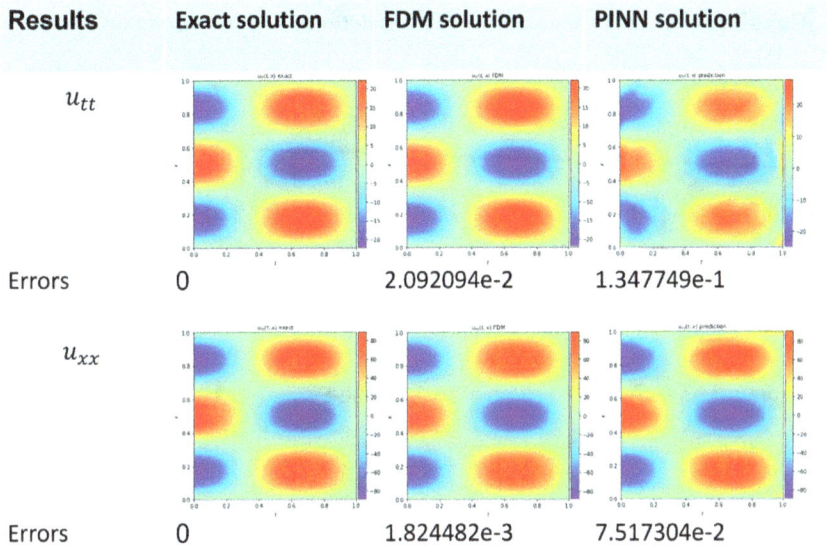

Figure 4.49. Comparison of FDM and PINN solution in a one-dimensional wave motion equation.

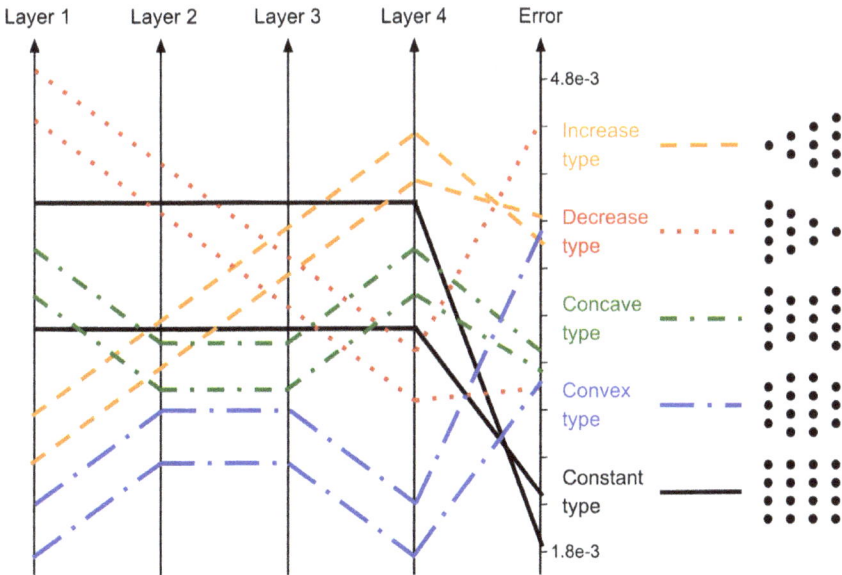

Figure 4.50. NN structure and E_{int}.

Results	Exact solution	FDM solution	PINN solution
u_{xx}			
Errors	0	5.242151e-2	5.955793e-2
u_{yy}			
Errors	0	5.830156e-5	1.189393e-1
u_{xy}			
Errors	0	6.524948e-4	2.910281e-2

Results	Exact solution	FDM solution	PINN solution
v_{xx}			
Errors	0	1.193476e-3	2.345729e-2
v_{yy}			
Errors	0	6.802550e-1	9.841444e-1
v_{xy}			
Errors	0	1.139795e-2	1.259519e-1

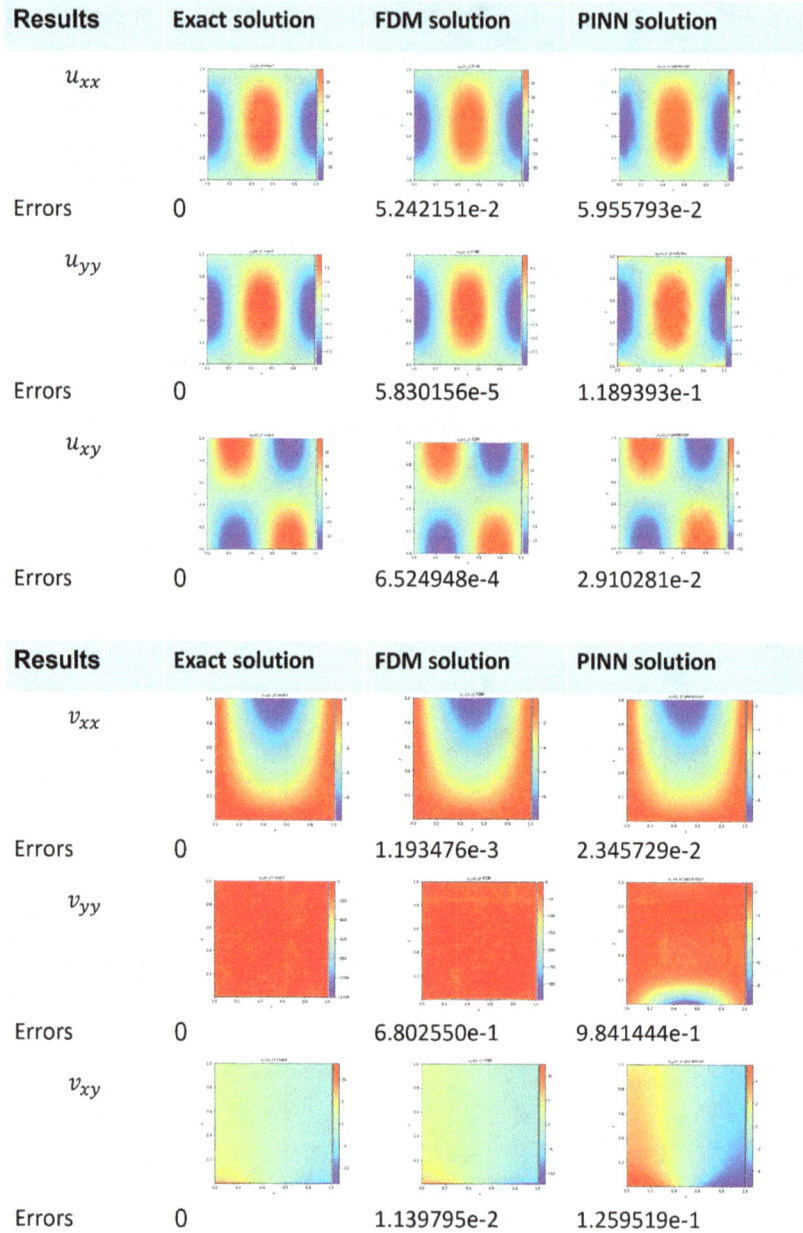

Figure 4.51. Comparison of FDM and PINN solutions in a two-dimensional stress equation.

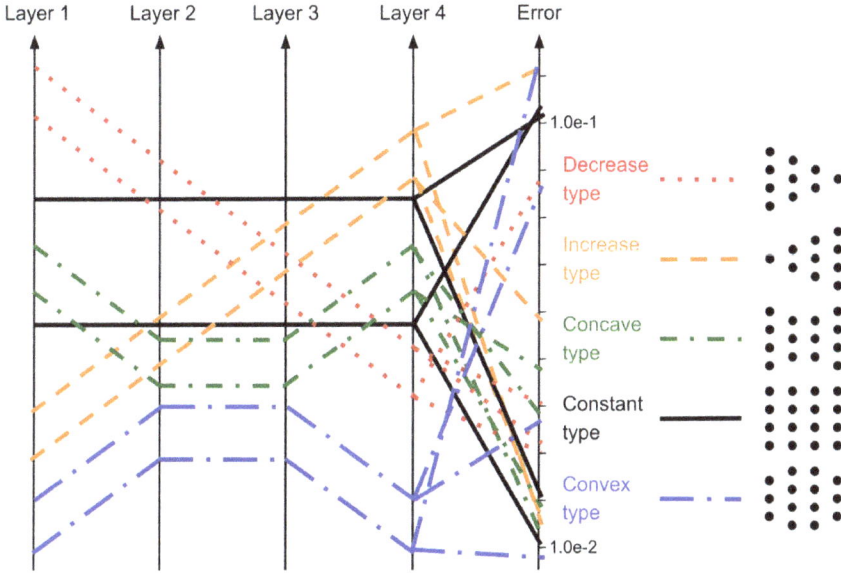

Figure 4.52. NN structure and E_{int}.

based on their exact solutions. For comparison, in FDM, Δx, Δt (mesh width) are determined so that the Courant number is 1 or less and the average is as large as possible within the domain. Figure 4.53 shows the results of comparison with FDM. Of these two methods, the PINNs method provides the most accurate solution (Figure 4.53).

Figure 4.54 shows the parallel coordinates of E_{int} when the approximate solution of the 3D AD equation is computed in N10.

4.3.5. *Implementation of the PDE solution method*

4.3.5.1. *Implementation using Colab*

The main processing flow of the code that implements the derivation of the 3D advection–diffusion equation described so far is as follows.

(1) Import the library and configure TensorFlow
(2) Define and initialize the PINN class

- Neural network weight and bias initialization
- TensorFlow session configuration and variable definition
- Setting placeholders for input and output variables

Results	Exact solution	FDM solution	PINN solution
u_t			
Errors	0	1.790425e-2	6.611525e-2
u_x			
Errors	0	4.487404e-2	2.905886e-2
u_y			
Errors	0	4.736407e-2	2.656753e-2
u_z			
Errors	0	5.043872e-2	2.296206e-2

Figure 4.53. Comparative results of FDM and PINN methods in a three-dimensional advection diffusion equation.

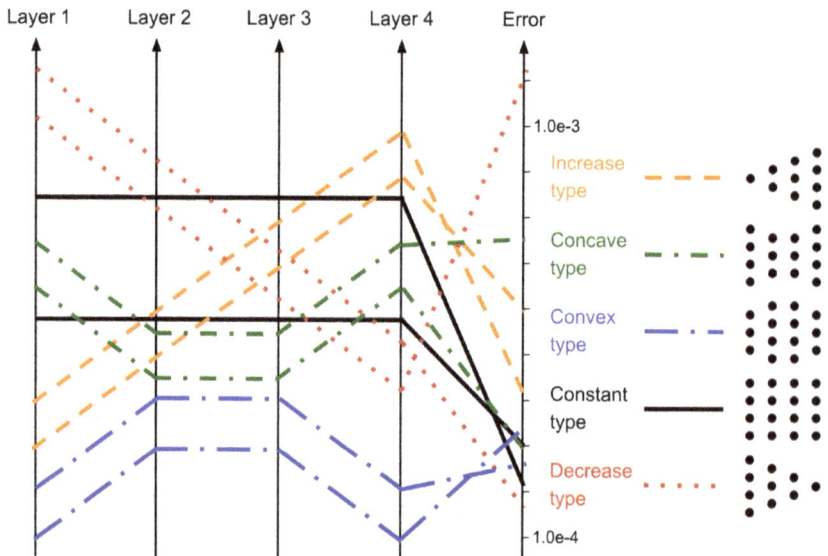

Figure 4.54. NN structure and E_{int}.

- Construction of neural network and prediction of output
- Definition of loss function and setting of optimization method

(3) Neural network training

- Update weights and biases by optimization method using training data
- Calculation and display of loss function values

(4) Forecast

- Input forecast data and get output forecast

Specifically, the inputs of the neural network are the spatiotemporal variables (x, y, z, t) and the outputs are the corresponding dependent variables (u). Neural networks are used to predict output variables from input variables. The loss function is the mean squared error (MSE) of the difference between the predicted output and the true output. The Adam (Adaptive Moment Estimation) optimizer is used as the optimization method and the weights and biases are updated for the specified number of iterations. Adam is a kind of optimization method using gradient descent. Adam is a technique that combines features such as learning rate tuning and momentum exploitation and is widely used in deep learning training.

The code for importing the library and configuring TensorFlow is shown in Figure 4.55.

```
import tensorflow.compat.v1 as tf
tf.disable_eager_execution()
# print(tf.__version__)

import numpy as np
import time
from scipy.interpolate import griddata

import plotly.graph_objects as go

self.sess = tf.Session(config=tf.ConfigProto(allow_soft_placement=True,
log_device_placement=True)) # tf placeholders and graph
```

Figure 4.55. Library import and TensorFlow settings.

```
class PhysicsInformedNN_extend:
  def __init__(self, X, u, layers):
    self.lb = X.m(0)
    self.ub = X.max(0)
    self.x = X[:,0:1]
    self.y = X[:,1:2]
    self.z = X[:,2:3]
    self.t = X[:,3:4]
    self.u = u
    self.layers = layers
    self.weights, self.biases = self.initialize_NN(layers)
    self.sess =
tf.Session(config=tf.ConfigProto(allow_soft_placement=True,
log_device_placement=True))
    self.x_tf = tf.placeholder(tf.float32, shape=[None, self.x.shape[1]])
    self.y_tf = tf.placeholder(tf.float32, shape=[None, self.y.shape[1]])
    self.z_tf = tf.placeholder(tf.float32, shape=[None, self.z.shape[1]])
    self.t_tf = tf.placeholder(tf.float32, shape=[None, self.t.shape[1]])
    self.u_tf = tf.placeholder(tf.float32, shape=[None, self.u.shape[1]])
    self.u_pred = self.net_u(self.x_tf, self.y_tf, self.z_tf, self.t_tf)
    self.u_t_pred, self.u_x_pred, self.u_xx_pred, self.u_xxx_pred,
self.u_y_pred, self.u_yy_pred, self.u_yyy_pred, self.u_z_pred,
self.u_zz_pred, self.u_zzz_pred = self.net_f(self.x_tf, self.y_tf, self.z_tf,
self.t_tf)
    self.loss = tf.reduce_mean(tf.square(self.u_tf - self.u_pred))
    self.optimizer_Adam = tf.train.AdamOptimizer()
    self.train_op_Adam = self.optimizer_Adam.minimize(self.loss)
    init = tf.global_variables_initializer()
    self.sess.run(init)
```

Figure 4.56. Defining and initializing the PINN class.

The code for defining and initializing the PINN class is shown in Figure 4.56.

This code defines two classes: the PhysicsInformedNN_extend class and the PhysicsInformedNN_minimum class.

The PhysicsInformedNN_extend class defines a neural network for deriving the 3D advection–diffusion equation. Specifically, the initialize_NN method initializes the weights and biases, and the net_u method computes the output of the neural network. Also, the net_f method computes the time and space derivatives of the output. As explained in Chapter 3, TensorFlow has a tf.GradientTape API for automatic differentiation, i.e., computing the gradient of the computational result with respect to the input variables. TensorFlow "records" on a "tape" all operations performed within the context of a tf.GradientTape. TensorFlow then uses that tape and the gradient associated with each operation recorded on it to compute the gradient of the "recorded" computation using top-down automatic differentiation (reverse mode). The PDE residuals are computed as follows (Figure 4.57):

The PhysicsInformedNN_minimum class similarly defines a neural network for solving 3D advection–diffusion equations, but omits some differentiation calculations.

These classes have a neural network trained in the train method and a prediction using the trained model in the predict method.

The code for the specific prediction part is as follows in the predict method (Figure 4.58):

```
def PDERE(U):
  return np.abs(U[1] - D*(U[3] + U[6] + U[9]) + Beta_x * U[2] + Beta_y *
U[5] + Beta_z * U[8]).mean()
```

Figure 4.57. PDE residual calculation.

```
tf_dict = {self.x_tf: X_star[:,0:1], self.y_tf: X_star[:,1:2], self.z_tf:
X_star[:,2:3], self.t_tf: X_star[:,3:4]}
u_star = self.sess.run(self.u_pred, tf_dict)
u_t_star, u_x_star, u_xx_star, u_xxx_star, u_y_star, u_yy_star,
u_yyy_star, u_z_star, u_zz_star, u_zzz_star = self.sess.run([self.u_t_pred,
self.u_x_pred, self.u_xx_pred, self.u_xxx_pred, self.u_y_pred,
self.u_yy_pred, self.u_yyy_pred, self.u_z_pred, self.u_zz_pred,
self.u_zzz_pred], tf_dict)
```

Figure 4.58. Obtaining predicted output values from input data.

```
fig = go.Figure(data=go.Isosurface(
   x=X_v.flatten(),
   y=Y_v.flatten(),
   z=Z_v.flatten(),
   value=values.flatten(),
   # isomin=10,
   # isomax=40,
   opacity=0.3,
   # caps=dict(x_show=False, y_show=False)
   ))
fig.show()
```

Figure 4.59. Visualization of output data.

The specific output data visualization code is as follows (Figure 4.59):

This code uses Plotly's library to create a 3D plot. Specifically, we are creating a diagram to display an isosurface. Let us look at the details of the code.

Line 1 uses go.Figure() to create a new figure. This diagram contains data.

The second line uses go.Isosurface() to create the isosurface data. This isosurface is the surface corresponding to a specific value in 3D space.

The following is a description of each parameter:

- $x = X\_v$.flatten(): Specifies the value of the X coordinate. X_v is a 3D data array, converted to a 1D array using the flatten() method.
- $y = Y\_v$.flatten(): Specifies the Y coordinate value. Y_v is similarly converted to a 1D array.
- $z = Z\_v$.flatten(): Specifies the value of the Z coordinate. Z_v is similarly converted to a 1D array.
- value $=$ values.flatten(): Specifies the value of the isosurface. The values have also been converted to a one-dimensional array. Based on this value, an isosurface is plotted.
- opacity $= 0.3$: Specifies the transparency of the isosurface. It can be specified in the range from 0 to 1. In this case, the isosurface will be 30% transparent.

- caps = dict(x_show = False, y_show = False): Sets not to show caps (end faces). This will hide the isosurface endfaces.

The last line uses fig.show() to display the figure.

4.3.5.2. *Implementation using FEMAP*

In stress analysis by FEM, a system has been commercialized in which the user defines the domain of interest, gives boundary conditions, performs analytical computations, and visualizes the results. This section implements stress analysis with PINNs using one such system, Finite Element Modeling and Post-Processing (FEMAP). FEMAP is a computer software system used for modeling finite element analysis (FEA) and visualizing the results. FEMAP provides an easy-to-use FEA modeling interface and is suitable for the analysis of complex structures [45]. FEMAP supports a wide variety of elements and material models and has advanced result visualization and analysis functions. Specifically, the commercial system FEMAP is used to enter shape data to create mesh. In FEM, a mesh is a geometric shape used to divide the domain to be analyzed into multiple smaller elements and analyze each element independently to obtain an overall picture. The mesh is usually made of triangular or rectangular elements, and the domain to be analyzed can be finely divided to increase accuracy. Figure 4.59 shows the details of the process (left side) and its user interface (right side). In stress analysis with PINNs, the following are required as a pre-treatment:

(1) Shape input, boundary conditions, and property settings
(2) NN hyperparameter setting and optimization
(3) Setting point group density and generation of selected point groups.

Among the above, FEMAP performs 1 and 3. In PINNs, the mesh itself is not needed, but the mesh points generated for the entered shape are used as the selected point group. Boundary conditions and property value settings can be used as is. Since FEMAP has an external file export function, a user selects the data needed for PINNs from the data exported using this function. For 2, the hyperparameters are optimized by setting them up in the system we developed. Since PINNs are used in this case, the FEMAP solution section (also called "solver") will not be used.

Stress analysis with PINNs outputs a displacement vector for each selected point, which can be transferred to FEMAP for visualization. The presumption in this case is that the mesh has already been defined in FEMAP. Specifically, the file is created as an addition to the data exported from FEMAP as a pre-process. FEMAP reads this file and performs post-processing. The post-processing consists of the following:

(4) NN model output
(5) NN model visualization

In the NN model output, data are exported in a form that FEMAP can process, and the NN model visualization uses FEMAP functions.

Chapter 5

Physically Based Surrogate Model

5.1. About the CAE surrogate model

PINNs are expected to play important roles in CAE surrogate model (SM) construction. CAE stands for "Computer-Aided Engineering" and refers to technologies and tools that utilize computers to support design, analysis, simulation, and other tasks related to engineering [45]. Surrogate means a substitute as in surrogate mother. In CAE, an SM is used when the results of interest cannot be computed efficiently. Most engineering design problems require numerical simulation to evaluate the objective function as a function of design variables. For example, to find the optimal wing shape for an aircraft, engineers compute the airflow around the wing for various shape variables such as length, curvature, and material. However, for many real-world problems, a single numerical simulation may take minutes, hours, or even days to complete. As a result, routine tasks such as design optimization, design space exploration, sensitivity analysis, and what-if analysis become difficult to perform because they require thousands or millions of numerical simulation evaluations. One way to reduce this burden is to use an approximate model called SM.

5.1.1. *Example of SM construction*

Two examples of SM construction that the author has performed are introduced in the following.

5.1.1.1. *Case 1: Heat management of laptop computers*

Temperature prediction for laptop computers is positioned as an essential technology to improve CPU performance and reduce the chassis size. The CPU package is represented as a thermal resistance network [47] consisting

T_1: Temperature of package

T_2: Temperature of lead wire

T_3: Temperature of silicon chip

T_4: Temperature of substrate

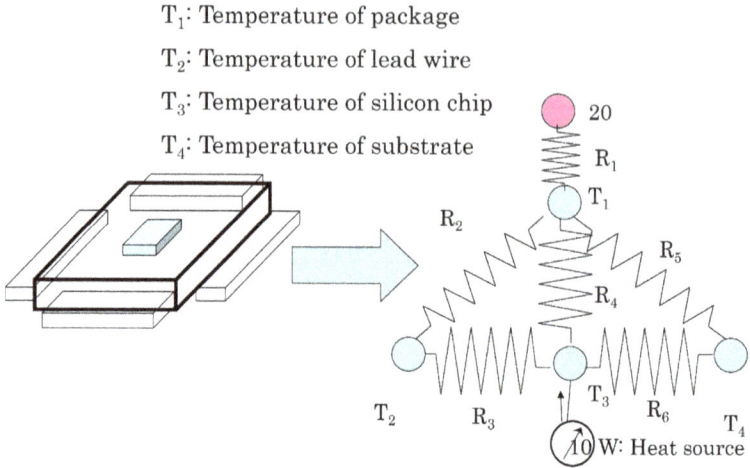

Figure 5.1. Thermal resistance network for CPU packages.

of multiple thermal resistances, and the simulation results for the CPU package and the space around it are used to identify their thermal resistance values (Figure 5.1).

Specifically, the computer chassis in which the CPU package might be placed is expressed as several boundary conditions, and a CAE model is computed for that number of boundary conditions. Using these results, the thermal resistance is computed using optimization methods. This thermal resistance network that takes into account various situations (called a compact model) is used when detailed CAE models are computationally time-consuming, which is considered a kind of SM.

5.1.1.2. *Case 2: Heat management at the switched reluctance motor (SRM)*

This section gives an example of temperature analysis in a switched reluctance motor (SRM). This motor is inexpensive and suitable for high-speed rotation because of its simple structure without permanent magnets and windings in the rotor. Thus, it is used in various applications such as home appliances (washing machines, vacuum cleaners), EV drives, and industrial applications. The rotor of the SRM is equipped with multiple protrusions to rotate using magnetic resistance. When the rotor rotates, the air flowing between the protrusions causes more wind loss than a normal cylindrical

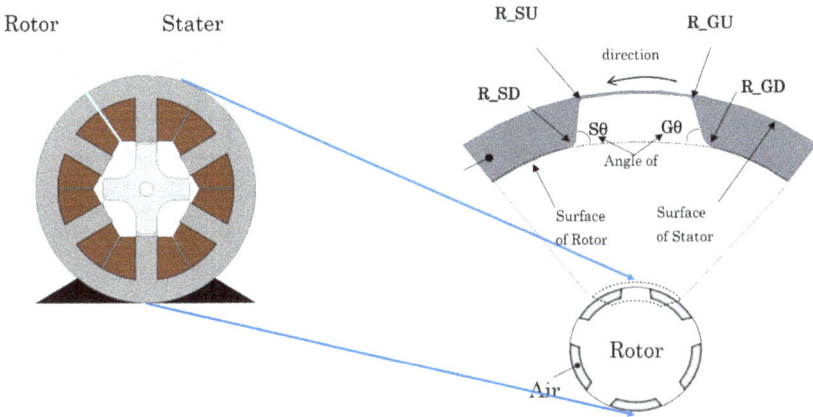

Figure 5.2. Structure of switched reluctance motor (SRM).

rotor, resulting in a rise in the average temperature of the rotor surface (Figure 5.2).

Three-dimensional thermo-fluid analysis software has been used to predict these temperatures. Until now, CAE analysis using this software required a lot of computation time to analyze a large number of designs (it took about 30 minutes for the computation of one design). With an SM, the computation time per design became only a few seconds, and the output was instantly provided. This allowed us to increase the number of designs from a few dozen cases to tens of thousands.

In this case, the following three types of regression analysis models using Excel analysis tools and solvers have been developed for the SM to derive temperature prediction equations that can instantly calculate temperature predictions for any point during the design phase:

- For Type 1, multiple regression analysis was performed using only the first-order terms of the six parameters as variables, using a total of seven coefficients including the parameters plus the intercept.
- For Type 2, a multiple regression analysis was performed with seven parameters $(S\theta, R\_SU * S\theta, G\theta, S\theta2, G\theta2, R\_GU * G\theta, G\theta * S\theta)$ selected from a total of 27 parameters: six parameters in the first-order terms (six), second-order terms (six), and the product terms in two variables (15), which have a large contribution to the rotor temperature increase.

Parameters		Minimum	Maximum	Number of levels
Angle of salient-pole rotor	*Sθ*	2	160	2(80)
	Gθ			
Top fillet	*R_SU*			
	R_GU			
Bottom fillet	*R_SD*	0.5	8	0.5(16)
	R_GD			

	Average error [%]	Maximum error[%]
Type1	1.7	8.9
Type2	1.3	8.6
Type3	0.6	4.2

Figure 5.3. Surrogate model for SRM temperature prediction.

- For Type 3, Sigmoid activation function and deep learning (one hidden layer) were used.

Of these, Type 3 was the most accurate (Figure 5.3).

5.1.1.3. *Use of Excel in SM construction*

Regression, in statistics, means fitting the model $Y = f(X)$ to data when Y is a continuous value. Another way to put it is to fit a model between the objective variable Y and the explanatory variable X on a continuous scale. If X is one-dimensional, it is called a simple regression, and if X is two-dimensional or more, it is called a multiple regression. If Y is discrete, it is called classification. Regression analysis involves analyzing data using regression, often creating the perception of employing a costly tool or developing a specialized program. This section shows that this is provided as one function of the commonly used Excel analysis tool.

As already discussed, Excel analysis tools support everything from correlation analysis to statistical tests and regression analysis, so the basic analytical techniques can be covered with the Analysis ToolPak. Excel Analysis ToolPak is one of Excel's add-ins that allows users to analyze data just by clicking or entering simple parameters. With the Analysis ToolPak, users never have to write special functions or codes. Since it is easy to install, even beginners can easily perform data analysis. In CAE analysis, objective variables, such as temperature and stress at a certain point, and design variables, such as geometry parameters, boundary conditions, and initial conditions, are defined in each column; a user provides enough number of rows for the number of designs. Then, an SM can be developed by running the regression analysis of the Excel Analysis ToolPak. The results

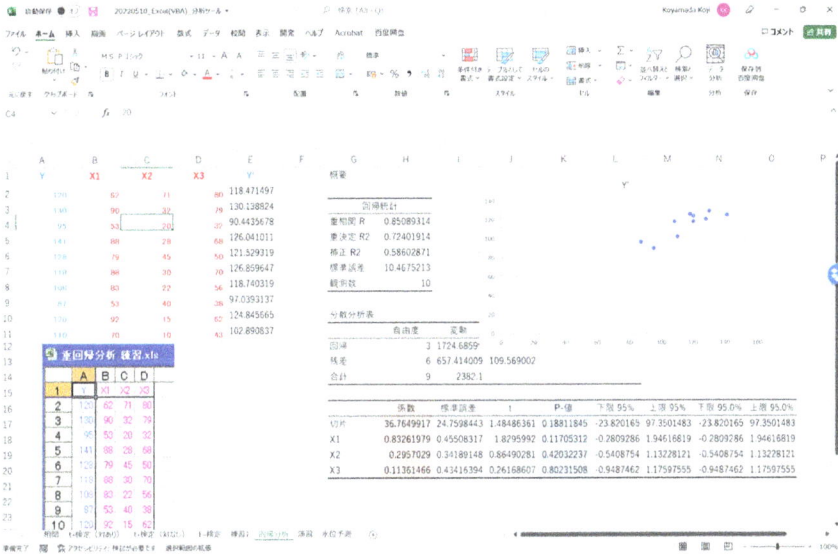

Figure 5.4. Regression analysis results using Excel analysis tools (three design variables).

of the regression analysis also show indicators for removing design variables that do not contribute much to the objective variable. Thus, it makes it easy to create compact models. Figure 5.4 shows the results of regression analysis using the Excel Analysis ToolPak (three design variables). Among the results, the cells where the coefficients for the design variables and their p-values are stored are filled in beige and blue.

In addition, the Excel Solver tool can be used to create SMs with even higher precision. With this function, users can perform regression analysis and NN modeling without programming. As already discussed, an Excel Solver is a function that sets the result of a formula calculation as a target value, specifies multiple constraints to obtain that result, and changes arbitrary cells to find the optimal value. To use Solver, users must enable the Solver add-in, as in the Analysis ToolPak. Using mathematical methods related to optimization, the maximum and minimum values, etc., are obtained within a specified range. To set the target value, cells in which the calculation results are output in the objective cell are specified, and the groups of cells whose values are to be varied are specified as variable cells. When the solution button is pressed after performing the above, optimal values are provided in the objective cells.

To perform a regression analysis using the Excel Solver, first calculate
the sum of the error squares of the difference between the predictions by the
regression and the simulation results on a sheet and designate the result
as the target cell. Next, the regression coefficients can be calculated by
specifying each coefficient of the regression analysis in the variable cells
and running Solver to minimize the objective cells. You can check to see if
this calculation is correct by creating the same model in the Excel Analysis
ToolPak and checking the results.

To create an NN model using the Excel Solver, an activation function
that performs a nonlinear transformation is applied to the output of the
regression analysis, which is used as a new output. This creates an NN
model for one layer. Figure 5.5 shows the NN model and its regression
analysis results created using Excel Solver. The explanatory variables are
varied on a regular grid, and their objective function values are visualized
three dimensionally.

Using the ExNN described in Chapter 2, Section 7, Paragraph 3, a
deep learning model can be constructed by specifying the number of layers
required and the number of neurons per layer. In this case, the target
cell is called the loss function in the NN model. The advantage of the
NN model is that it satisfies the universal approximation theory from a

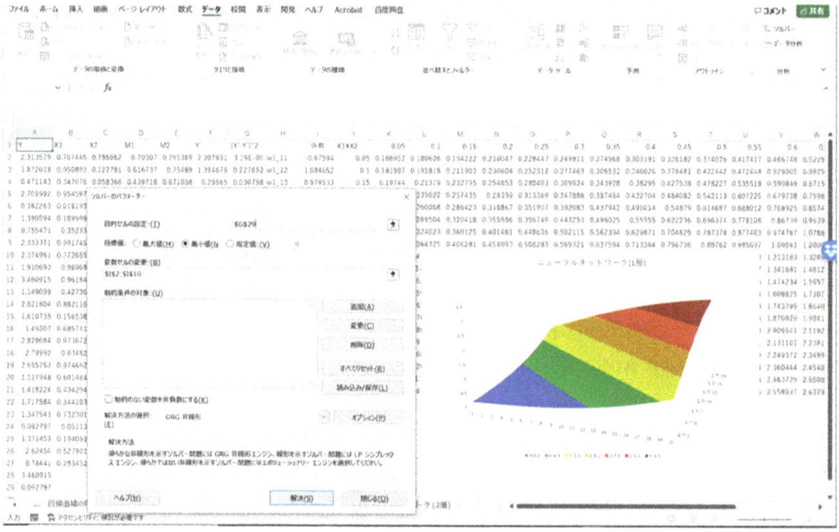

Figure 5.5. The NN model created with Excel Solver and the results of the regression
analysis.

mathematical point of view. As already mentioned, this is the theorem where "an NN consisting of one hidden layer with a finite number of neurons can approximate an arbitrary continuous function under some conditions". This is the background theory that has led to NNs having many application fields.

5.1.2. *Expectations for physics-based learning*

Some CAE engineers are concerned about the SM in terms of its accuracy. This is because even if the NN is highly accurate from a mathematical standpoint, there is no way to be sure that it is physically correct. For CAE models that CAE engineers are interested in, there is a corresponding PDE; thus, a technique that uses physics-based learning to achieve even higher accuracy, or PINNs, has been proposed. Physics-based learning uses automatic differentiation to first compute arbitrary partial differential values for an NN model to compute PDE residuals. Here, automatic differentiation refers to a technique for efficiently and accurately computing functions represented by computer programs. Next, by adding the PDE residuals to the loss function of the NN model, this technique builds an NN model with physically improved accuracy. Its range of applications is expanding to PDE derivation and finding solutions from big data. It is not impossible to perform the same task using Excel alone. But Excel would require the use of a deep learning framework such as Tensorflow to efficiently compute the partial derivatives of the objective variable for the design variables.

5.2. Application to carbon neutrality (CN)

The SM is expected to be used for CN policy formulation based on scientific findings. It is because what-if analyses of computational science simulation models based on environmental data measured by citizens play an important role in realizing evidence-based policymaking (EBPM) by citizens. The EBPM is a practice that uses rational basis (evidence) for clearly defined policy objectives, rather than relying on ad hoc episodes for policy planning [49]. The promotion of the EBPM using data such as computational simulations, which have important relevance for measuring the effects of policies related to carbon neutrality, will enhance the effectiveness of policies and contribute to securing citizens' trust in government.

For example, the use of heat-shielding pavements with enhanced solar reflective properties, rooftop greening, wall greening, and other measures has been realized to control the heat island phenomena. Simulation based on computational science is expected to play an important role in assessing the effectiveness of such heat-shielding pavements, but it is not yet available for citizens to use. One solution to this issue is a highly anticipated physics-based surrogate model for the general public with no experience or knowledge of computational simulation.

Once it becomes available, even without experience or knowledge of computational simulation, any citizen interested in policymaking can be assumed to have some idea of what-if analysis. In the case of a heat problem, for example, when many citizens are interested in the radiative properties of construction materials and their impact on the temperature, experts run computational simulations with different radiation characteristics and use deep learning to create models that explain temperatures by radiation characteristics. This model can be classified as a surrogate model because it allows temperatures to be evaluated from radiative properties without the need for actual computational simulations. Using the surrogate model, citizens interested in heat issues can perform a what-if analysis. The surrogate model improves accuracy from a statistical point of view, but this result does not guarantee physical validity.

To ensure physical validity, PINNs may be employed for surrogate models. In PINNs, environmental data are measured to derive physical models such as PDEs, and then citizens and other interested parties solve the PDEs using various environmental parameters. In PINNs, space-time coordinates are fed into the input layer, and a loss function is constructed from the calculation results at the output layer and the physical data associated with its coordinates. Next, an automatic differentiation operation is performed on the NN model to add PDE residuals and regularization terms to the loss function. This operation allows the construction of a surrogate model with guaranteed physical validity.

This physics-based surrogate model can be used to assist in generating evidence to formulate environmental policy. Policymaking begins with clarifying issues, and what-if analysis is essential in this phase to recognize the problem and to estimate the factors that affect the problem. While environmental data such as temperatures and heat indices are what we focus on in the issue of global warming, identifying the factors that influence them requires examination of physics as well as intuition and experience. Surrogate models based on physics play an important role because not many policymaking officials specialize in physics. In order to utilize surrogate models with physical validity in policymaking, it is necessary to properly incorporate physics into surrogate models.

No research has been proposed on techniques to improve the accuracy of surrogate models using PINNs. In previous surrogate modeling, predictable locations were predetermined.

For example, in Section 1, Paragraph 1, a numerical fluid analysis model of a rotor in a switched reluctance motor (SRM) was created, and a surrogate model that predicts rotor temperature using varying design parameters was explained. In temperature prediction, the surrogate model was able to derive a temperature surrogate model that could instantly compute with an error of less than 4.2%. Meanwhile, the ability to predict temperature in areas other than rotors remains to be achieved in the future. Another company has developed a finite element model that predicts the strength of burst plates in air conditioning compressors, and a physics-based surrogate model has been proposed to predict the strength of burst plates when design parameters are varied. Strength prediction by this surrogate model was found to be more than 1,000 times faster than performing a physics-based analysis. But the assurance of physical validity in addition to flexibility in specifying the location of strength prediction remains a future challenge.

Physics-based surrogate modeling technology using PINNs, which will be developed through this research, is expected to be implemented in society. Specific examples are described in the following. Once the effectiveness of the physics-based surrogate modeling technique is confirmed for the simulation of the thermal environment of a specific city block, which requires a huge amount of computation time, it is expected that the technique will contribute to global warming mitigation and adaptation measures at the city, national, and global scales. Global warming countermeasures can be broadly classified into mitigation, which is the reduction of the emissions of greenhouse gases, the causal substances of global warming, and adaptation, which is to control adverse effects of global warming by adjusting natural ecosystems and social and economic systems in response to climate change. Mitigation measures control the effects on natural and human systems in general by controlling atmospheric greenhouse gas concentrations. On the other hand, adaptation measures are characterized by directly controlling the effects of global warming on specific systems.

This section focuses on the verification of the effectiveness of radiant cooling materials to control heat problems as an example [48]. As the heat from the sun reaches the earth, the earth radiates the heat back into space, but some of the heat is absorbed by the greenhouse gases, increasing the energy of the gases and causing temperatures to rise. This radiative material attempts to release radiant heat from the earth into space through the atmosphere as a window and is expected to be an adaptive measure for global warming [50].

The surrogate model described here can be used for what-if analysis, which is important in policymaking. Specifically, it allows for instant visualization of reliable results regarding where and what kind of heat control measures will reduce average summer temperatures. It can also instantly visualize changes in the possibility of levee breaches in relation to river levels. In order to effectively visualize data using surrogate models in policymaking meetings, it is necessary to design a user interface that is easy for citizens to use. In designing the user interface, citizens are interviewed in advance to determine what it should look like.

Chapter 6

Closing Remarks

This book used specific examples to explain the methods for analyzing and visualizing physical data using NNs. The first part covered the spatially discrete data in which time is irrelevant and explained the following:

(1) Page information from ancient literature data captured with a three-dimensional CT machine.
(2) Explanation of a method for visualizing plasma regions with high accuracy from a magnetic field group computed from electromagnetic field analysis results in fusion reactors. Then, the following were explained using the physical data:

    (1) Methods using NN to derive a PDE that explains its physical data.
    (2) How to solve PDE using NN.

We hope that this book will provide readers with hints for solving their problems. Now, we would like to conclude this book by suggesting some future directions for research in this area.

In this book, we described how to use NN to find solutions to PDEs from physical data given as boundary conditions. Based on the definition given at the beginning, physical data consist of the space-time coordinates in which the data are defined, and the physical data associated with those coordinates. The PDE solution method is now in the process of being compared with conventional methods (FDM, FEM, PM, etc.) to determine how they are grouped. But considering the merits of using NN models, we believe that the future direction is to apply this method to surrogate models, which have recently been utilized more in computational engineering.

In computational engineering, surrogate models are used when the results of interest cannot be easily computed [46]. Most engineering design problems require numerical simulation to evaluate the objective function as a function of design variables. For example, to find the optimal wing shape for an aircraft, engineers compute the airflow around the wing for various shape variables such as length, curvature, and material. However, for many real-world problems, a single numerical simulation may take minutes, hours, or even days to complete. As a result, routine tasks such as design optimization, design space exploration, sensitivity analysis, and what-if analysis become difficult to perform because they require thousands or millions of numerical simulation evaluations. One way to reduce this burden is to build an approximate model, called a surrogate model. In the last few years, surrogate models using NNs have been widely used. But these are simple approximate models that do not refer to the PDE that describes the target numerical simulation, and there are sometimes concerns about ensuring physical accuracy. To find a future direction for the surrogate model, one needs to consider how to deepen its relationship with PDEs.

Suppose now that a numerical simulation is run with various boundary conditions, physical data in the region of interest are obtained, boundary conditions are entered into the NN input layer, and physical data are

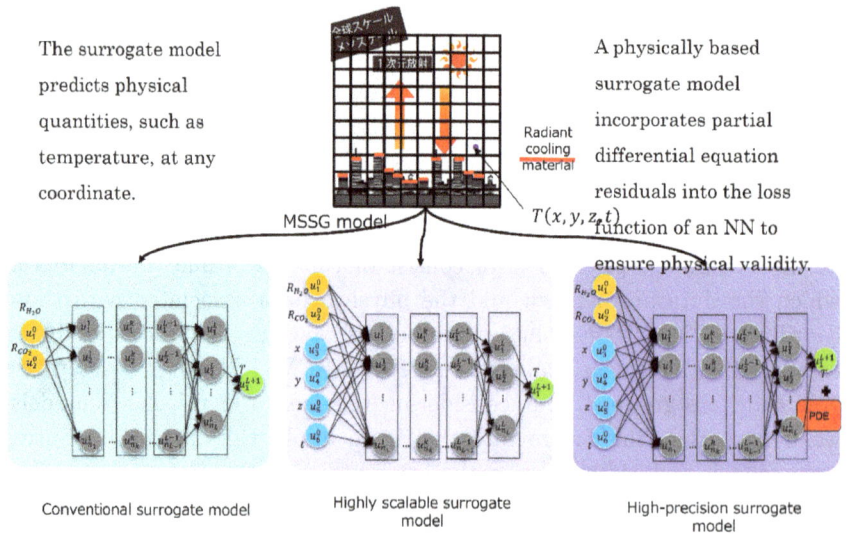

Figure 6.1. Physics-based surrogate model.

also entered into the NN output layer to construct an NN model. This NN model will be able to predict the physical data when unknown boundary conditions are given, without being affected by PDEs that explain the numerical simulation. This is the conventional surrogate model commonly used today (see lower left of Figure 6.1). The NN, which can be created by inputting the space-time coordinates that acquire physical data besides boundary conditions, can predict physical data from arbitrary space-time coordinates and boundary conditions. This is called a highly scalable surrogate model (see the lower part of Figure 6.1). It is surely more scalable, but physical validity remains untouched.

The question we are asking here is whether we can incorporate PDE residuals in the process of building a surrogate model in order to improve the physical validity of the surrogate model? There are several answers to this question, but using PINNs is a sure way to achieve this. The NN model created by making it learn multiple boundary conditions can provide a place where arbitrary boundary condition data can be specified, and physical validity can also be guaranteed, resulting in a so-called highly accurate surrogate model (see lower right of Figure 6.1).

References

[1] Brunton SL, Proctor JL and Kutz JN. (2016). Discovering governing equations from data by Sparse identification of nonlinear dynamical systems. *Proceedings of the National Academy of Sciences*, **113**(15), 3932–3937.

[2] Google, Welcome to colaboratory. Google Colab, https://colab.research.google.com/notebooks/intro.ipynb.

[3] Abadi M, *et al.* (2016). TensorFlow: Large-scale machine learning on heterogeneous systems. Software available from tensorflow.org.

[4] Gulli A and Pal S. (2017). *Deep Learning with Keras: Implementing Deep Learning Models and Neural Networks with the Power of Python*. Packt Publishing Ltd.

[5] Harris CR, *et al.* (2020). Array programming with NumPy. *Nature*, **585**(7825), 357–362.

[6] Tosi S. (2018). *Matplotlib for Python Developers: Effective Techniques for Data Visualization with Python*. Packt Publishing Ltd.

[7] Healy K. (2018). *Data Visualization in Python with Seaborn*. O'Reilly Media.

[8] Monjezi S. (2020). *Python Plotly: A Comprehensive Guide*. Packt Publishing Ltd.

[9] Juniper J. (2020). *Exploring Data with Python and Jupyter Notebook: An Illustrated Guide*. O'Reilly Media.

[10] Qi CR, *et al.* (2017) PointNet: Deep learning on point sets for 3D classification and segmentation. In *Conference on Computer Vision and Pattern Recognition*, arXiv:1612.00593.

[11] Wu C, Wang J and He H. (2019). Origin-enhanced cationic surfactants enable efficient DNA compaction and dynamic DNA nanotechnology. *Langmuir*, **35**(5), 1656–1665.

[12] Hagenauer T. (2019). *Learning Grafana 6*. Packt Publishing Ltd.

[13] Craig AB. (2013). *Understanding Augmented Reality: Concepts and Applications*. Morgan Kaufmann.

[14] Pfister H, Lorensen B, Bajaj C, Kindlmann G, Schroeder W, Avila L and Machiraju R. (2006). The Transfer Function Bake-off. *IEEE Computer Graphics and Applications*, **26**(1), 16–22.

[15] Sakamoto N, *et al.* (2010). Improvement of particle-based volume rendering for visualizing irregular volume data sets. *Computers & Graphics*, **34**(1), 34–42.

[16]   Montgomery DC, Peck EA and Vining GG. (2012). *Introduction to Linear Regression Analysis*. John Wiley & Sons.
[17]   Breusch TS and Pagan AR. (1979). A simple test for heteroscedasticity and random coefficient variation. *Econometrica*, **47**(5), 1287–1294.
[18]   Kennedy J. (1995). The stepwise regression procedure: Some controversies and recommendations. *Journal of Educational and Behavioral Statistics*, **20**(2), 101–129.
[19]   Hamilton JD. (1994). *Time Series Analysis*. Princeton University Press.
[20]   Lasdon LS, Waren AD, Jain A and Ratner M. (1978). Design and testing of a generalized reduced gradient code for nonlinear programming. *ACM Transactions on Mathematical Software (TOMS)*, **4**(1), 34–50.
[21]   Dantzig GB. (1963). *Linear Programming and Extensions*. Princeton University Press.
[22]   Coello CAC, Veldhuizen DA and Lamont GB. (2007). *Evolutionary Algorithms for Solving Multi-Objective Problems*. Springer.
[23]   Cybenko G. (1989). Approximation by superpositions of a sigmoidal function. *Mathematics of Control, Signals and Systems*, **2**, 303–314.
[24]   Baydin AG, *et al.* (2018). Automatic differentiation in machine learning: A survey. *Journal of Machine Learning Research*, **18**, 1–43.
[25]   Koyamada K, *et al.* (2021). Data-driven derivation of partial differential equations using neural network model. *International Journal of Modeling, Simulation, and Scientific Computing*, **12**(2), 2140001_1–2140001_19.
[26]   Mizuno S, Koyamada K and Natsukawa H. (2018). Derivation of partial differential equations. Japan Society for Simulation Technology, AI Research Committee (Japanese).
[27]   Josh B and Lipson H. (2007). Automated reverse engineering of nonlinear dynamical systems. *Proceedings of the National Academy of Sciences*, **104**(24), 9943–9948.
[28]   Guo LZ, Billings SA and Coca D. (2010). Identification of partial differential equation models for a class of multiscale spatio-temporal dynamical systems. *International Journal of Control*, **83**(1), 40–48.
[29]   Lingzhong G and Billings SA. (2006). Identification of partial differential equation models for continuous spatio-temporal dynamical systems. *IEEE Transactions on Circuits and Systems II: Express Briefs*, **53**(8), 657–661.
[30]   Maziar R, Perdikaris P and Karniadakis GE. (2019). Physics-informed neural networks: A deep learning framework for solving forward and inverse problems involving nonlinear partial differential equations. *Journal of Computational Physics*, **378**, 686–707.
[31]   Hao X, Chang H and Zhang D. (2019). DL-PDE: Deep-learning based data-driven discovery of partial differential equations from discrete and noisy data. http://arxiv.org/abs/1908.04463.
[32]   Rudy SH, Brunton SL, Proctor JL and Kutz JN. (2017). Data-driven discovery of partial differential equations. *Science Advances*, **3**(4), e1602614.
[33]   Maziar R and Karniadakis GE. (2018). Hidden physics models: Machine learning of nonlinear partial differential equations. *Journal of Computational Physics*, **357**, 125–141.

[34] Koyamada K. (2019). Effects of initial condition comprehensiveness on derivation accuracy of partial differential equations from big data. Oukan Symposium (Japanese).

[35] Patankar SV. (1980). *Numerical Heat Transfer and Fluid Flow*. CRC Press.

[36] Whitham GB. (1974). *Linear and Nonlinear Waves*. John Wiley & Sons.

[37] Burgers JM. (1948). A mathematical model illustrating the theory of turbulence. *Advances in Applied Mechanics*, **1**, 171–199.

[38] Hackbusch W. (1994). *Elliptic Differential Equations: Theory and Numerical Treatment*. Springer.

[39] Carslaw HS and Jaeger JC. (1959). *Conduction of Heat in Solids* (2nd edn.). Oxford University Press.

[40] Duff GFD and Bacon DW (1997). *Waves in Fluids*. Cambridge University Press.

[41] Zienkiewicz OC, Taylor RL and Zhu JZ. (2013). *The Finite Element Method: Its Basis and Fundamentals* (7th edn.). Butterworth-Heinemann.

[42] Haberman R. (2004). *Applied Partial Differential Equations: With Fourier Series and Boundary Value Problems* (5th edn.). Pearson.

[43] Rao SS. (2017). *Applied Finite Difference Methods*. John Wiley & Sons.

[44] Belytschko T, *et al.* (1996). Meshless methods: An overview and recent developments. *Computer Methods in Applied Mechanics and Engineering*, **139**(1–4), 3–47.

[45] Bi Z. (2018). Chapter 1 — Overview of Finite Element Analysis. In *Finite Element Analysis Applications*. Academic Press, 1–29.

[46] Liang L, *et al.* (2018). A deep learning approach to estimate stress distribution: A fast and accurate surrogate of finite-element analysis. *Journal of the Royal Society Interface*, **15**(138), 20170844.

[47] Specht E, *et al.* (2016). Simplified mathematical model for calculating global warming through anthropogenic CO_2. *International Journal of Thermal Sciences*, **102**, 1–8.

[48] Chen Z, *et al.* (2020). Radiative cooling: Principles, progress, and potentials. *Advanced Materials*, **32**, 43.

[49] Nutley SM, Walter I and Davies HTO. (2007). Chapter 1 — Using evidence — introducing the issues. In *Using Evidence: How Research Can Inform Public Services*. Policy Press, 10–18.

[50] Jeremy NM. (2019). Tackling climate change through radiative cooling. *Joule*, **3**(9), 2057–2060.

[51] Hu K, *et al.* (2021). Visualization of plasma shape in the LHD-type helical fusion reactor, FFHR, by a deep learning technique. *Journal of Visualization*, **24**(6), 1141–1154.

[52] Han Z, *et al.* (2023). High-precision page information extraction from 3D scanned booklets using physics-informed neural network. *Journal of Visualization*, **26**(2), 335–349.

Index

www.ingramcontent.com/pod-product-compliance
Lightning Source LLC
Chambersburg PA
CBHW050558190326
41458CB00007B/2095